AMAZON BESTSELLING AUTHOR

Brian Harris

WHY

Our Search for Meaning in the 21st Century

© 2019 by Brian Harris

All rights reserved. No part of this work may be reproduced or transmitted in any form or by any means, electronic or mechanical, including photocopying and recording, or by any information storage or retrieval system, without permission in writing from the publisher.

ISBN 9781790698349

CGS Communications, Inc.
www.cgscommunications.com

CONTENTS

Introduction   5.

1. Fake News   13.

2. Why do some people blame God for pain and suffering?   35.

3. What does 21st century learning teach us about pain and suffering?   57.

4. Are pain and suffering different for humans?   75.

5. Is there life after death?   91.

6. Why do some people resist change?   113.

7. Will we be able to create eternal life in the 21st century?   131.

8. Will virtual reality allow us to create our own narrative?   169.

9. Three degrees Fahrenheit   189.

10. Survival of the fittest   217.

11. Why are we here?   241.

12. Some final thoughts   271.

Notes   287.

Suggested Readings   323.

"The secret to life is meaningless unless you discover it yourself."
W. Somerset Maugham,
Of Human Bondage

INTRODUCTION

Four teenagers are celebrating the end of the school year when their car spins off a road late at night and banks into a tree. The four teenagers are killed in this horrendous accident. A phone call alerts me that my help as a crisis counselor is required in assisting the family members and friends impacted by this terrible tragedy.

I have spent over three decades working as a professional counsellor. Within this framework, I also worked as a member of a Tragic Events Response Team for a significant period of time. As a counselor on this team, I was often asked to assist people whose family members or friends had experienced tragedies. These misfortunes included drug overdoses, terminal diseases, suicides, car accidents, workplace accidents, assaults, shootings, and even murders. I have also spent a significant amount of time counseling adults who have lost their jobs which can be a significant form of loss and working with children (and adults) who have experienced family conflicts (including divorce which is another form of loss).

In situations like these, a common question often surfaces. The question is "Why?" although there are various forms of this that include:

- *Why did this happen?*
- *Why am I suffering?*
- *Why is God doing this to me?*
- *Why did God allow this to happen?*
- *Why am I being punished?*
  and so on....

It could be added that at the root of these questions are three of the most basic questions that humans have likely sought to answer since the beginning of our time on this planet.

6 / WHY

> *- Why are we here?*
> *- Why do pain and suffering exist?*
> *- Is there life after death?*

Without appropriate answers to these questions, people may struggle to live happy, meaningful lives. Complicating one's quest to find answers to these questions is the fact that some of the answers offered in previous centuries may no longer be applicable in the 21st century. As Douglas Adams writes in his popular science fiction satire, *The Hitchhiker's Guide to the Galaxy*:

> *"This planet has—or rather had—*
> *a problem, which was this:*
> *most of the people living on it*
> *were unhappy for pretty much*
> *most of the time."*[1]

Peace, Prosperity, and More Suicides

Although the 21st century has not brought the world wars that were experienced in the 20th century or the magnitude of disease and conflict that were prevalent in previous times throughout history, many people are still unhappy and live unfulfilling lives.

For example, in 2016 approximately 64,000 people died from drug overdoses in the U.S.[2] To put this into perspective, this is more deaths than the number of U.S. soldiers killed in both the Vietnam and Iraq wars combined. Add to this the fact that 42,773 people died from suicide in the U.S. in 2014. It is reported that suicide in the United States has surged to its highest levels in nearly 30 years.[3] Also consider the unfortunate fact that 13,286 people died from gun violence in 2015 in the U.S.[4] When one adds these numbers together and considers that these deaths do **not** account for the actual number of people addicted to drugs, those who attempted suicide, or those who were involved in some form of gun violence, then clearly there is significant unhappiness and disillusionment in a country such as the U.S. (and throughout this book I will share information that shows that the U.S. does not have the sole ownership of these problems).

On the other hand, some experts have argued that we have entered a period of peace and prosperity that the world has never experienced before in its history. If we were to look back in time, we would find that a thousand years ago life expectancy for a woman was about 25 years.[5] At the current time there are at least 50 countries in the world where the life expectancy for a woman is over 80 years.[6] In the past 10 centuries life expectancy for women has more than tripled in many countries.

In his book *The Better Angels of Our Nature*, Steven Pinker states:

> *"Believe it or not—and I know that most people do not—violence has declined over long stretches of time, and today we may be living in the most peaceable era in our species' existence."*[7]

In a January, 2018, article in *Time Magazine,* Pinker also said:

> *"According to the latest data, people are living longer and becoming healthier, better fed, richer, smarter, safer, and more connected—and at the same time, even gloomier about the state of the world."*[8]

We could very well ask how is it at a time of greater prosperity, peace, and longevity that so many people could be gloomier?

Are Religions Losing Their Significance?

Perhaps it's time to reexamine the answers that have been traditionally given to those important questions that begin with why.

For thousands of years, most humans have found their answers to the questions about life's meaning in religion. At a time when more and more people are turning away from organized traditional religions, perhaps there is a need to look at the why questions without blindly using the church or even God to provide the default answers.

In September, 2016, Barna Research released the following statement:

> *"The Christian church has been a cornerstone of American life for centuries, but much has changed*

in the last 30 years. Americans are attending church less, and more people are experiencing and practicing their faith outside of its four walls. Millennials in particular are coming of age at a time of great skepticism and cynicism towards institutions—particularly the church."[9]

In a related article, Barna found that two-thirds of Christians faced doubt in their beliefs (and this is coming from people who actually identify themselves as subscribing to basic Christian beliefs).[10] A 2014 study from the Pew Research Center found that:

"The share of Americans who say they are 'absolutely' certain God exists has dropped sharply, from 71% in 2007 to 63% in 2014. And the percentages who say they pray every day, attend religious services regularly and consider religion to be very important in their lives also have ticked down by small but statistically significant margins."[11]

Even the evangelicals who have wielded political power in the U.S. recently are not immune to a reduction of its membership. A recent article in *Newsweek* stated:

"Only 10% of Americans under 30 identify as white evangelicals. The exodus of youth is so swift that demographers now predict that evangelicals will likely cease to be a major political force in presidential elections by 2024."[12]

These trends are not just associated with the Christian church. According to a 2012 Pew Research Study, "more and more young people in America are claiming no religious affiliation. Muslims living in American are not immune to this phenomenon."[13]

While some will look at the previous statistics and conclude that the problem of disillusioned people could be resolved by a return back to the church, it could also be argued that organized religion as it currently exists no longer has relevant answers to the why questions for many people, and this is one of the major reasons why so many people have turned away from it. Clinging to myths and traditions from the past, instead of being more aware of 21st century learning, may also have something to do with so many people being unhappy. If the religious beliefs of our parents and grandparents no longer bring meaning in the 21st century, then where does one turn to find lasting happiness?

God as a Jealous, Petty, Genocidal, Vindictive, Homophobic, Misogynistic, and Racist Tyrant

It's also possible that more and more people are not as blind as past generations have been in looking at God as he is presented in Holy Scriptures.

The God of the Old Testament which is the basis for Judaism, Christianity, and Islam (the world's three major religions, although it could be argued that Judaism is not one of the world's major religions in terms of membership, but I have included it here because it provides the foundation for both Christianity and Islam) is often described in the Bible as being jealous, petty, genocidal, vindictive, homophobic, misogynistic, and racist. He is a God who tested one of his devoted followers—Abraham, who is considered to be the founding father of the world's three major religions—by asking him to sacrifice his son (Genesis 21-22). As Abraham places his son on an altar and the fire is about to begin, God changes his mind (consider the psychological trauma this boy suffered). In today's world God and Abraham would be charged with child abuse.

The Bible is full of stories of God expressing his fury in a manner that is repulsive and sometimes beyond belief. When Moses reportedly led the Jews out of Egypt, God ordered that the Israelites should kill the firstborn child of every Egyptian before they left (Exodus 12:29). Is this the God that people in the 21st century want to worship? In another instance God tells the Israelites to "go and smite Amalek, and utterly destroy all that they have, and spare them not, but slay both men and women, infant and suckling, ox and sheep, camel and ass" (1st Samuel 15:2-3). This represents the extermination of a race. This is known as genocide. Is this the God that people in the 21st century want to worship?

In another Biblical recounting, God commanded the Israelites to attack another group of people known as the Midianites and to kill every male, even the male children, and to kill every woman. Then God tells them that they can save any female children to keep for themselves to rape or keep as slaves (Numbers 31:17-18). Is this the God that people in the 21st century want to worship?

Of course, some will respond that God as portrayed in the New Testament is much gentler and more loving, but this is still the same God as described in the Old Testament—his image has just been sanitized for the most part to be more acceptable to a different generation and culture. In spite of God appearing to be more approachable and kinder in the New Testament, we are also told that he will someday unleash the apocalypse, a time of great suffering that will result in billions of people being sentenced to an eternal fiery hell. How can a loving God be so ruthless? Even Jesus, who is portrayed as being the ultimate act of God's love to humans, said (as recorded in the New Testament), "If any man comes to me and hates not his father, and mother, and wife, and children, and brethren, and sisters, yea and his own life also, he cannot be my disciple" (Luke 14:26). Jesus appears to be telling people that in order to be one of his followers they have to be willing to reject their own family, and even be willing to give up their own lives. Isn't this what cults do? Isn't this a demand made by terrorist extremists in recruiting new members? Is this the God who people want to worship in the 21st century?

Why is it that so much to do with God in the Bible portrays him as being jealous and vengeful? Is God inherently violent, or is this portrayal the result of ancient beliefs that focused on a god to be feared? Could the 21st century usher in a new image of God?

Some might argue that in the 21st century fewer people believe in the literal account of scriptural stories, so do they really matter anymore? A 2017 Gallop poll showed that:

> *"24% of Americans still believe that the Bible is to be taken literally, word for word while 26% believe that the Bible is a book of fables, legends, history and moral precepts recorded by man."*[14]

This means that about half of Americans are somewhere in the middle of this divide, believing that some parts should be taken literally and some parts should be taken metaphorically, but even if only 24% of people believe in a literal interpretation of the Bible, this number is enough to cause problems in our world as I will show in later chapters.

At a time when the character and contributions of so many historical figures are under the moral microscope, God and any Holy Scriptures are not going to be exempt from such examination.

I grew up as a member of an evangelical church (and this was a mainstream internationally accepted church) that hugely shaped and controlled my thinking about the world around me. From a very early age I was taught to blindly accept the church's answers to the questions about finding meaning in life. I attended church two to three times on Sunday, and during the week there was rarely an evening when one our family members wasn't at our church for some function. Eventually I became a youth leader within the church and the teacher of adult Bible study groups. After three decades of very active involvement in this church, I became disillusioned with many of its teachings and I left, seeking life's meaning from other sources.

In this book, at times I will share my story, a story that strips away the falsehoods of religion and helped me find the truth. I will also share the stories of many people I have worked with in my role as a counselor as they tried to remove the chains of past learning to find meaning in the 21st century.

The Intent of this Book

The intent of this book is to examine how 21st century learning is challenging some long-accepted beliefs related to our search for meaning. I will offer some insights on how 21st century learning can help you to find meaning based on fact rather than superstition. Perhaps such realities might provide more meaningful answers in your search for meaning.

In looking at our search for meaning in the 21st century, we need to take into account the rapid and dramatic changes that are occurring in technology and science. The Internet in a recognizable form appeared in 1990 (it first emerged for a select number of people in 1983). Personal computers were first introduced in 1975 as kits for the do-it-yourself hobbyist. The first draft of the human genome didn't occur until 2001. In the past 50 years, the growth of technological advances has been incredible. We are now on the threshold of being able to create life and prolong it. We are on the threshold of technological advances that are staggering, but we are also on the threshold of creating weapons of mass destruction that the world has never faced before. And then, of course there's climate change.

With these rapid changes will come new ways of thinking. New ways of answering life's difficult questions.

What does 21st century learning teach us about answering some of the why questions?

Let's begin our journey.

- 1 -
Fake News

During this past summer, a solar eclipse occurred in a path across North America. In the area where I live, I observed about a 70% eclipse. Watching locations on TV where they experienced a 100% eclipse, I found the experience moving. In spite of being aware of the scientific explanation for the eclipse, the actual eclipse somehow, forgive the pun, eclipsed the physical reality of what was happening.

The Evolution of God
As I thought further about the eclipse, I couldn't help but think about our earliest ancestors' reactions to events such as an eclipse, a storm, a drought, or any other natural occurring phenomenon. It's easy to understand how the first humans hundreds of thousands of years ago perceived some supernatural force (the gods) controlling these natural occurrences. Over many, many years these fundamental perceptions of the world led to beliefs that became deeply rooted in the psyche of humans.

In considering our earliest ancestors, it's not difficult to understand how a random bolt of lightning (especially if it struck a family member or caused a forest fire that destroyed their homes) was interpreted as a punishment from some unseen power such as a god.

It's estimated that the first Homo sapiens, the classification of humans that we belong to, appeared more than 300,000 years ago[1] (and our ancestors known as Australopithecus—a genus of hominins—occurred almost 4,000,000 years ago).[2] Is it possible that the underlying superstitions of our ancestors, who had no scientific knowledge to explain natural disasters that occurred around them, have been passed down to us?

Millions of years passed between the appearance of our ancestors and the establishment of a preeminent ancient civilization such as the Egyptians where we have concrete historical records detailing their traditions and religions starting in approximately 3100 BCE until its conquest by Alexander the Great in 332 BCE.[3] Although we don't have written historical records of our earliest ancestors (we do though have some interesting archaeological evidence that we will look at in a few more pages), we have strong historical evidence of the beliefs of ancient civilizations such as the Egyptians. We might ask how these ancient beliefs still impact our search for meaning in the 21st century?

The Egyptians believed that their gods "meted out punishment to individuals for wrongdoing."[4] It's very likely that the humans who lived many years before the Egyptians had similar beliefs. After all, common sense would tell us that the formal religion of the Egyptians, which included sacrificing and elaborate funeral practices, didn't just appear out of thin air. And along the same lines, the religions of today must have borrowed some of their beliefs from the pagan religions of the past. In his book *The Evolution of God*, Robert Wright states:

> *"...the idea is that 'primitive' religion broadly, as recorded by anthropologists and other visitors, can give us some idea of the ancestral milieu of modern religions."*[5]

Although logic (and archaeological evidence) would tell us that a civilization such as the Egyptians borrowed some religious ideas and practices (such as sacrificing to the gods) from previous cultures, for the moment let's forget about what happened before the Egyptians. A study of the ancient Egyptian civilization clearly shows that both animal and sometimes even human sacrifices to their gods was a part of their religious practices.

Sacrificing to a god was a way to curry favor with a god for many ancient civilizations such as the Egyptians. It is based on the belief that the gods could intervene in their lives, a belief that is a key component of most major religions today (some 5,000 years later) even though the act of sacrificing an animal or another human has almost completely disappeared.

How many people in a crisis situation even today still attempt to bargain with God? The bargaining usually goes something like this: "If you heal me (or insert the name of a loved one in here), I'll attend church every Sunday forever." Bargaining with God to gain a favor has its roots in ancient civilizations thousands of years ago (and perhaps much, much further back into pre-history). For ancient civilizations, sacrifices were thought to be a way to sustain their gods. Individuals in ancient societies interacted with their gods through daily prayers, hymns and private offerings.[6]

Similar beliefs were echoed by the Greeks and Romans, and many other cultures of that era, who attempted to please their gods with sacrifices and various forms of praise. For example, in Greek literature, especially in Homer's epics, the sacrifice of animals is used to gain the favor of the gods. In the *Odyssey*, Homer describes a huge ceremony where some 4,500 people offer 81 bulls in sacrifice to the god Poseidon.[7]

For people who still believe that there is a personal God who interacts with them daily, this belief evolved from ancient pagan religions thousands of years ago. We will explore this thought further in the next chapter.

Does God Punish People?

Although few people today believe in Egyptian or Greek gods lashing out at them for their failings (does anyone really believe that Zeus is standing on a mountain top firing lightning bolts at us), most major world religions have incorporated the philosophy that "God punishes us when we do wrong" into their teachings.

This belief has become so pervasive in most cultures, regardless of religious beliefs, that God often becomes the scapegoat for tragic events on both a worldwide and personal scale. I hope to show in this book that such thinking is inherently false and that such thinking can actually interfere with finding meaning and achieving happiness. In the next chapter we will examine this belief much more closely.

Will Our Beliefs Change in the 21st Century?

Our beliefs are generally the result of cultural traditions, influential leaders, teachers, Holy Scriptures, revelations from science, and even myths from the past. For the most part, the originators of this wisdom were inspired by their observations and understanding of the world around them although some would argue that various Holy Scriptures were inspired directly by God, and as a result should be the definitive basis of meaning (even if some of these "inspired" writings may have been more applicable at the time when they were written, or they have proved to be factually incorrect, or they are repulsive, or they are contradictory to so-called inspired writings from other religions).

In the same way that science and technology are evolving in the 21st century, perhaps there is a need for religion and philosophy to evolve as well.

Harvard physicist and bestselling author Lisa Randall writes:

> *"Science certainly is not the static statement of universal laws we all hear about in elementary school. Nor is it a set of arbitrary rules. Science is an evolving body of knowledge. Many of the ideas we are currently investigating will prove to be wrong or incomplete."*[8]

If science is willing to evolve in its search for the truth, then doesn't it make sense that the answers to our search for meaning should be evolving as well?

Randall argues that new knowledge does not necessarily result in the discarding of old rules or ways of thinking. Sometimes the evolution of knowledge can expand or even enhance what we previously believed, although in some cases new knowledge will challenge us to reconsider how some aspects of our past might be preventing us from finding happiness and fulfillment in the present.

Even though we may have a different understanding of our world in the 21st century than our ancestors had hundreds or even thousands of years ago, there are still many traditions and practices (and ways of thinking) that we follow today that have their roots far in the past. If any of these traditions or practices are interfering with our current pursuit of happiness, perhaps it's time to reconsider their place in our lives.

Adam and Eve and Cancer

As a youth, at the age of 12 or 13, I remember squirming through a sermon at church where the minister ranted on about the perils of sin which was a fairly typical sermon at our evangelical church, along with hell and eternal damnation. Church leaders have known for many centuries that fear is the most powerful way to control their congregations (and keep the money flowing into the church).

The minister spoke tenaciously about how sin first began in the Garden of Eden (regularly pounding the pulpit with his fist, the blows reverberating through the microphone attached to the pulpit as though someone had shot a gun). As his voice continued to rise, he talked about how Adam and Eve had brought pain and suffering into the world. I felt uneasy with what he was saying. I had recently learned that a boy in one of my classes at school was dying of cancer.

I couldn't imagine how Adam and Eve could have anything to do with the suffering this boy was experiencing, especially when my strong interest in science told me that the story of Adam and Eve couldn't possibly be taken literally. From my readings, even at this early age, I knew that there was no fossil record to support the concept of the independent creation of two humans who were the parents of the human race, not to mention the problems of the in-breeding that would have occurred as a result of the incestuous relationships that would have taken place if humanity started with only two people.

If you believe in Adam and Eve, then you must also believe that that their children had sex with each other, and for generations it was a regular occurrence for brothers, sisters and cousins and perhaps even their parents to all have sex with each other. Sorry, Bible supporters, but this didn't make sense to me when I was a kid and as an adult, it makes even less sense.

Getting back to my story, after the service, I approached the minister and asked him how he would explain the story of Adam and Eve in light of the archaeological evidence that supported the evolution of humans, rather than the spontaneous creation of Adam and Eve. His answer was quick and firm. He said, "Son, never question the word of God. God demands complete obedience." And of course, by implication, never question the word of a church leader because he also demands complete obedience.

The Problem of Complete Obedience to the Word of God

In the book *The God Delusion*, the author Richard Dawkins, an ethologist and evolutionary biologist, states:

> *"Christianity, just as much as Islam, teaches children that unquestioned faith is a virtue. You don't have to make the case for what you believe. If somebody announces that it is part of his faith, the rest of society, whether of the same, or another, or of none, is obliged, by ingrained custom, to respect it without question; respect it until the day it manifests itself in a horrible massacre like the destruction of the World Trade Center, or the London or Madrid bombings."*[9]

Throughout history millions have died and many more have suffered because people were taught by religious leaders to never question the word of God (or perhaps it would be more correct to say to never question their interpretation of the word of God). Countless wars have been caused by religious beliefs.

Interestingly in the First and Second World Wars, a number of soldiers on both sides stopped fighting on Christmas Day and even engaged with each other in a friendly manner to celebrate Christmas, only to resume their brutal killing the next day because each side believed their God was with them (even though in theory they were all worshipping the same God).[10]

In the First and Second World Wars, by and large both sides of the conflict believed in the same Holy Scriptures. Both sides believed that God was with them in their fight. How wrong could they have been?

Blind obedience to the so-called word of God can lead to unnecessary suffering and unhappiness.

There are additional problems related to blind obedience to the word of God when we consider that within specific religions there are a variety of interpretations related to the "inspired word" of God.

For example, in the Christian religion (and we could also include Muslims and Jews in this conversation) we find that some Christians are vehemently opposed to homosexuality while other Christians are more accepting. For those Christians who are opposed to homosexuality, they often quote verses found in the Bible that make statements such as, "If a man has sexual relations with a man as one does with a woman, both of them have done what is detestable. They are to be put to death; their blood will be on their own heads" (Leviticus 20:13).

The same people who quote verses from the Bible to support their views against the LGBTQ community generally ignore other commands in the Bible, such as the punishment of death for any children who curse their parents (Exodus 21:17), or the stoning of rebellious teens (Deuteronomy 21:20-21). For those Christians who accept members of the LGBTQ community, their acceptance is based on selected verses from the Bible that talk about loving and accepting others without judging them (Matthew 7:1-5).

While many Christians who live in North America might say they are more accepting of people from the LGBTQ community, I read in this morning's newspaper about a local woman who was kicked out of her evangelical church because she was gay. The article stated:

> *"After disclosing her relationship [with members of the church], Mills [the woman involved] said that she had several conversations with church leaders and members. They wanted her to repent her sexual orientation, to be 'restored' to the Bible's teaching."*[11]

The Bible doesn't have sole possession of such extreme thinking. In the Quran, Muslims can find justification for the beheading of non-believers (8:12-13), or the amputating of the hands of someone who has stolen something from another person (5:38), two practices that are still performed in some countries of the world. It is reported that the Quran has more than 100 verses that talk about waging war on nonbelievers.

Some might argue that the verses in the Old Testament of the

Bible, or the Quran that relate to violence or the abhorrent treatment of women, that we would consider to be troubling today, should be viewed in a historical context and are not necessarily applicable today. Unfortunately one only needs to jump to the New Testament (of the Christian Bible) where God has supposedly had a change of face to become a God of love and equality to find verses that still state man's superiority over women, as evidenced in verses such as "Wives, respect and obey your husbands" (1 Peter 3:1) or "Wives should obey their husbands in everything" (Ephesians 5:22-24).

Of course, we also have the end times as outlined in the book of Revelation in the New Testament where we read about the wrath and judgment of God being poured out on earth (Revelation 16:1-21) including plagues of sores, poisonous blood in the seas, more blood for drinking water, fiery scoring heat, a great earthquake and hailstorm. And then after this, we have eternal punishment in a fiery hell for non-believers where they will be tormented forever (Revelation 20: 10, 14-15). And this from a loving God?

Is this the God we want to give our complete obedience to?

As author, physician, and oncologic pathologist Ali A. Rizvi writes:

> *"What creator of nebulae, time-distorting black holes, and hundreds of billions of stars and galaxies would be so insecure to consider it an absolute requirement that the inhabitants of this infinitesimally tiny speck of a planet regularly kiss his ass in gratitude for the infinitesimally tiny fragment of time they're here, or suffer an eternity of fire and torture?"*[12]

Why Did He Wait So Long?

It's estimated that our universe is about 14.5 billion years old and that our planet is about 4.5 billion years old. And it's estimated that Homo sapiens (that's us) are a few hundred thousand years old. For all those who believe that God created us to worship him, have you ever wondered why he waited around 10 billion years after creating our universe before creating our planet, and then he waited another 4 billion years before creating humans? If humans are so important to God, why did he wait for more than 14 billion years before we came into existence?

Perhaps we need to take a closer look at the roots (and facts) of some of our beliefs if we are to find happiness and fulfillment in the 21st century. Is it possible that our religious leaders have been leading us in the wrong direction? Is it even possible that writers of Holy Scriptures purposely created a terrifying image of a ruthless God because fear was a great way to keep the masses obedient?

Is it possible that the basis of our world religions is the result of myths that have been told over and over so many times that we now accept them as facts. As Dr. Adam Rutherford writes in his bestselling book *A Brief History of Everyone Who Ever Lived*:

> *"There is no contemporary evidence even for the existence of Jesus Christ, arguably the most influential man in history. Most of our tales about his life were written in the decades after his death by people who had never met him. Today, we would seriously question that, if it were presented as historical evidence. Even the accounts that Christianity rely on, the Gospels, are inconsistent and have irreversibly mutated over time."*[13]

Christmas: False Facts?

Let's look at an event that most readers would be familiar with—Christmas.

Let's start with the date, that of December 25th. There is nothing in the Bible, the early writings of Christian leaders, or any writers of history of that era that tell us that Christ was born on December 25th. The first written mention of the birthdate of Christ appears to have come in the 3rd century when Clement of Alexandria raised this question. Clement actually proposed seven potential dates, none of which were December 25th.

It wasn't until 440 A.D (more than 400 years after the birth of Christ) that the Christian church officially proclaimed December 25th as being the date of Christ's birth.

It is generally believed that December 25th was borrowed from the Romans who celebrated the reemergence of the rebirth of their sun god, an event known as Saturnalia, during the winter solstice.

The Romans borrowed this date from the ancient Sumerians as well as the Egyptians. The Egyptians believed that their winter solstice celebration marked the rebirth of their god Horus.

Moving back to the Romans, December 25th was originally the celebration of the sun god (solis invicti nati—day of the birth of the unconquered sun), so our first fact related to the celebration of Christmas is that the original celebration and date had nothing at all to do with Christ, but likely had their roots in ancient pagan religions.

The Bible tells us (and the words of Christmas carols support this) that on the night of Christ's birth, the shepherds were out watching their flocks at night when an angel appeared to them to announce the birth of Christ. The only problem here is that sheep were corralled in the colder winter months in the geographical location where Christ was reportedly born. If we were to take the Bible verse literally about the shepherds tending their flocks on the night that Jesus was born, this would suggest that Christ's birth did not occur in December, but rather during a warmer season when the shepherds might actually be outside looking after their flocks.

The lore of Christmas tells us that three wise men followed a star that stopped over Bethlehem where Christ was born. First of all, the Bible doesn't tell us that there were three wise men and it doesn't refer to them as kings as sung about in the Christmas carol (and does anyone ever wonder as they are singing such a song that three kings coming on camel from the Orient would take months, and perhaps years, to actually arrive in Bethlehem?). Tracing the chronological record surrounding Christ's birth as presented in the Bible, we would find that the wise men, if they really existed, must have visited Christ months or even a year or two after his birth, but most certainly not on the night he was born. Could all those nativity plays and manger scenes be so wrong?

At a time, 2,000 years ago, when there were people who studied the night skies to observe stars and other celestial phenomena, there is no historical record anywhere (other than the Bible) of a bright star suddenly appearing on December 25th during the year when Christ was said to have been born.

What about Christ being born in a stable? The Bible doesn't actually say he was born in a stable (or a barn), nor does it say anything about farm animals being present at his birth, although barnyard animals are always found in today's nativity scenes and in some of the carols that are sung.

The Greek word that is used in the Bible to describe Christ's birthplace is kataluma which could be translated as "inn." Typically, any guests would stay in the upper rooms of such an inn. The Bible tells us that there were no rooms in the inn for Mary and Joseph (the parents of Jesus), but it does not tell us that they then went to a stable or barn. Having no room in the inn could simply be interpreted as there being no rooms in the upper level. When the upper level priority rooms in an inn were full, it was common practice for visitors to stay in the main room on the lower floor. This does not imply that Christ was born in a stable, although hundreds of years of Christmas sermons have focused on his humble beginning in a stable amongst barnyard animals.

Another central belief of Christmas is that Jesus was born of the virgin Mary, and that God impregnated her. Virgin births have been a central part of mythology for thousands of years. The ancient Egyptians (and you could add the Romans, Greeks, and even the Aztecs to this) believed that their gods regularly impregnated human virgins to give birth to new gods. Could the virgin birth simply be an attempt by early Christians to give greater significance to the birth of Christ by following the mythological tradition of a god impregnating a woman to give birth to a new god?

Then of course there's the record in Matthew 2 of King Herod ordering the killing of all male children two years old or under in the vicinity of Bethlehem in an attempt to kill baby Jesus. There is no historical record of any such mass murder (and how could this be missed?), even though the Catholic Church has identified these phantom murdered children as its first Christian martyrs (still celebrated even today as Holy Innocents Day on December 28th). In regards to this story, classical historian Michael Grant has said,

"The tale is not history, but myth or folk-lore."[14]

If we take away the date of December 25th, the stable, the wise men, the star in the sky, the virgin birth, and the barnyard animals, what do we have left? While it makes all the sense in the world for Christians to celebrate the birth of Christ, the actual celebration as it currently exists—from nativity scenes to carols—is permeated with false facts.

When myths and false facts reside in a story as important as the Christmas story, the story itself loses its significance. It then becomes easy to add additional myths such as Santa Claus to the narrative. In South Korea, Christmas is largely celebrated as a secular lovers' holiday, similar to Valentine's Day, especially among the young. If Christians would like to celebrate the true meaning of Christmas, one might argue that there is a need to separate the facts from the myths. If any religion doesn't care about the facts behind its beliefs and traditions, then it will struggle to survive in the 21st century.

As bestselling author Yuval Noah Harari writes in his book *21 Lessons for the 21st Century:*

> *"When a thousand people believe some made-up story for one month, that's fake news. When a billion people believe it for a thousand years, that's religion, and we are admonished not to call it "fake news" in order not to hurt the feelings of the faithful (or incur their wrath)."*[15]

Our Need For Stories

As humans, we don't have a very good track record of caring for our planet or each other in spite of billions of humans who believe in the God of ancient texts such as the Bible or Quran. How would our world change if our beliefs in the 21st century were based on facts instead of myths? How might our world be different if we could take the best of religious thought from past centuries and eliminate the idea that any one religion could somehow be superior to any other religion?

As humans, we appear to need stories. In the 21st century, which stories might best contribute to our current happiness and future survival? Is it possible that some of the new stories that will bring comfort and meaning into our lives might be based on fact rather than fiction, or at least focus on peace rather than violence, or acceptance rather than division, or human insight rather than divine inspiration?

Changing Facts, Changing Beliefs

Some archaeologists and scientists have stated that religion did not truly arise in early mankind until language had progressed to the point where religious concepts could be discussed. Philip Liebermann, a cognitive scientist at Brown University, states "human religious thought and moral sense clearly rest on a cognitive-linguistic base."[16] If that cognitive-linguistic base is still evolving in the 21st century, will this result in new answers to old questions?

Ralph Waldo Emerson stated that the "religion of one age is the literary entertainment of the next."[17]

In the early 1600s when Galileo proposed that it was the earth that rotated around the sun (a concept first proposed by the Greek astronomer Aristarchus of Samos almost 2,000 years earlier, and confirmed by Nicolaus Copernicus in the early 1500s),[18] his theory upset a belief system that had likely been in effect since the beginning of the human race.

Galileo's discoveries about the planets and the heavens were directly related to his improvement of the telescope that had been invented in the Netherlands by Hans Lippershey in 1608 (or Lippershey was at least the first person to file a patent on a telescope).[19] Galileo's observation of the world around him was altered by his use of technology.

Even though the Catholic Church ordered Galileo not to hold, teach or defend the Copernican theory that the earth rotated around the sun in any way whatever, either orally or in writing, Galileo continued in his research and writings.[20] Today we would think it ridiculous to even consider that the sun rotates around the earth. In the example of Galileo, the Catholic Church in their position against him ended up supporting a position of "false news" for hundreds of years in spite of scientific proof that they were wrong.

We are living at a time when false or fake news has often confused the general public as to what is factual and what is an outright lie. The next time a major story occurs somewhere in the world (especially if the story somehow involves American politics or religion), switch back and forth between CNN and FOX news. Even though the content of the story might be the same, the political orientation of the two news networks often filters the content in such a way that you might think you were watching two completely different stories. For one of the two networks (and sometimes possibly even both), their interpretation of some news is false.

Lies or Alternative Facts?

Fake news is nothing new. What is new is the ease (and speed) at which the Internet can spread such lies (and unlike some who would like us to believe that lies are just really "alternative facts," let's call a lie for what it is—a lie!). As we explore answers to the why questions, it's important to sift out any past forms of fake news that have contributed erroneously to current thinking related to our search for meaning.

In his book *1984*, George Orwell wrote:

> "The Ministry of Peace concerns itself with war, the Ministry of Truth with lies, the Ministry of Love with torture, and the Ministry of Plenty with starvation. These contradictions are not accidental, nor do they result from ordinary hypocrisy: they are deliberate exercises in doublethink."[21]

Fake News in History

Since the beginning of recorded history, doublethink and fake news have existed (which of course also suggests that fake history must have sometimes occurred).

For example, imagine being a scribe for an Egyptian Pharaoh. You have been asked to write a story about the Pharaoh on the interior walls of his burial chamber (inside a massive pyramid). In such a situation, what you write is completely biased by the fact that your life will be terminated (in some very brutal form of torture) if you don't please the Pharaoh. There is no way that you are going to note that the Pharaoh might have abused his subjects or that he might have squandered the nation's income on his lavish palace or tomb. No, you're going to embellish the life and accomplishments of the Pharaoh.

History is full of examples of misinformation. For example, in Trent, Italy, on Easter Sunday, 1475, a 2½-year-old child named Simonino went missing. Bernardino da Feltre, a Franciscan priest claimed that the Jewish community had murdered the child. Even though his account was untrue, the entire Jewish community in the city was arrested and tortured. Fifteen members of the Jewish community faced the cruel death of being burned at the stake. Throughout nearby towns, other similar acts of cruelty were inflicted on Jewish people even though the story that started the persecution was false.[22]

As the story continued unabated, the Christians in these communities spread fake stories about Jews murdering children and drinking their blood. It is often thought by historians that this macabre retelling has contributed to anti-Semitism since that time. And interestingly, Simonino—the missing child—was canonized by the Pope and is still considered to be a martyred saint by the Catholic Church (of course, the church never considered an apology to the Jewish people for the mistake that was made).

Another example of fake news occurred on August 25, 1835, (and the story continued for another five days) when the *New York Sun* ran a story about an English astronomer who had discovered life on the moon (through the use of a giant telescope). The article even contained descriptions of bizarre creatures living on the moon (such as two-legged beavers and man-bats). The story spread throughout the world before the newspaper admitted that the story was a hoax. The following is an excerpt from the article.

> *"...they were like human beings. They averaged about four feet in height, were covered, except on the face, with short and glossy covered hair, and had wings composed of a thin membrane, lying snugly upon their backs. The face, which was of a yellowish color, was a slight improvement upon that of the large orang outang."*[23]

There are people who believe that the holocaust never happened in spite of firsthand reports, photographs and authentic documented records. Some people believe that the American moon landing in 1969 was staged. There are even those who still believe that the earth is flat.

While we live at a time when there is often a very careful checking of the facts before a story is reported in the media, this is not necessarily an accurate description of past history (and even with all this checking today, incorrect facts are still sometimes reported, as are the twisting of the facts to suit the writer's biases).

If you had been hired by Hitler, Napoleon, Stalin, or some Popes (or just about any authoritarian ruler throughout history), you either wrote what your boss wanted you to write, or you didn't have your job anymore (and likely you no longer had your life either). As such, fake news has existed throughout history and as a result, our beliefs are sometimes based on falsehoods, rather than the truth. If we are to find true meaning for our existence in the 21st century, doesn't it make sense that we attempt to find our answers based on facts rather than fabrications?

It is reported that the current U.S. president (Trump) has lied or made false statements 4,229 times in less than his first two years as president.[24] This becomes even more disturbing when we consider that Trump's evangelical Christian base continues to wholeheartedly support him in spite of his constant false claims. Are conservative Christians really that gullible, or do they simply choose to ignore facts?

Not only does fake news exist because of the desire of a writer to please his leader (or boss) or to support his own biases, it also exists because it was a common practice for centuries for historians to write accounts of past events without actually interviewing anyone who might have been involved in the event. Often the writings occurred many years after the event which sometimes meant that there was no one still alive who had even been an actual witness to the event.

Herodotus, a Greek from the 5th century BCE is often considered to be one of our first historians. In describing the lengthy war between the Greeks and Persians, Herodotus drew on a number of sources, although he also incorporated rumors into his writings.[25] At the time, incorporating rumors was an accepted practice. It was also an accepted practice to write books years (and even generations) after the events occurred. Important books (take the Bible as an example) often had stories that were written many years (at least 30 years in the case of the New Testament) after the events occurred without any actual eye witnesses being interviewed.

With the most modern recording equipment, events that are occurring around the world right now sometimes get misinterpreted after a day or two. How much more potential is there for a story to change if it was written a hundred years after it apparently occurred, and there were no longer any firsthand accounts of what actually happened, and if the writer was biased to present a certain philosophical or religious point of view?

Does anyone really think there was a journalist or a historian sitting in the Garden of Eden to accurately record what transpired between Adam, Eve, and God? And while many people consider the story of Adam and Eve to be a metaphor, there are millions who view this story as the literal inspired word of God in spite of the absence of any fossil records which might support the idea that God inserted a man and woman into our world without them progressing through the evolutionary process that every other creature experienced.

Now you might be saying to yourself, does fake news really matter? Who really cares whether an ancient historian such as Herodotus didn't fact check his sources? Does it really matter if one news station reports the news with a completely different bias than another? As I will attempt to show in this book, it most certainly matters.

Fake News, Wars, and Persecution

Throughout history, millions of people have died cruel deaths in wars or have suffered various forms of persecution that were directly related to religious beliefs. What if some of these religions were based on false facts? What if some of the religious leaders cherry-picked verses from their accepted Holy Scriptures to justify their attacks on others?

On August, 23, 2008, Amanda Lindhout, a Canadian Journalist, was kidnapped by Islamist insurgents in southern Somalia. Kept captive, often chained in nightmarish living conditions for 15 months, Lindhout was repeatedly raped and abused by her devout Muslim capturers. In an attempt to stop the abuse, Lindhout attempted to convert to the Muslim religion by praying with her captors and by reading the Quran, but the abuse continued.

When Lindhout was able, on occasion, to talk to her captors about the way they treated her, she was told that their religion allowed them to do whatever they wanted with an infidel (a non-believer). As long as they prayed the required number of times each day, they could rape a defenseless woman because their religion allowed it.

When Lindhout talked with these men about their persecution of innocent people or even their willingness to die as suicide bombers, she found their preoccupation with the afterlife superseded any sense of personal responsibility for their actions here on earth. From their leaders, these often uneducated young men, had been brainwashed to believe that their religion was superior to others, and that non-believers should be killed (and they could do whatever they wanted with non-believing women).

As a result of her fifteen-month firsthand experience with Islamic fundamentalists during her captivity, Lindhout wrote:

> *"One thing about Islam is that paradise always beckons. Life is orientated towards the afterlife. Whatever pleasures you miss out on in this world, whatever comfort or richness or beauty is absent from your days and years, you will find it upon entering paradise, where pain, grit, and war disappear all together. Paradise is a vast perfect garden."*[26]

We might very well ask how many people of any religious orientation fail to take personal responsibility for the way they live, or even fail to enjoy each day of life to the fullest, because they are caught up in thinking about the paradise that is yet to come. What if this paradise doesn't actually exist? In Chapter 5 we will explore more on life after death in terms of 21st century learning.

In considering our lives in the 21st century, is it possible that millions of people (and it would be technically more correct to say billions of people) on our planet are allowing unsubstantiated beliefs from the past to determine how they seek happiness in the present?

If you had a serious medical problem that required surgery, would you rather be operated on by a modern physician with the most up-to-date technological equipment based on scientific advances, or would you prefer to be operated on by an ancient Egyptian physician with a knife (no anesthetic) and a book of spells along with magic potions? Most people would choose the modern-day doctor to help them. Why then is it that so many people choose rituals and traditions from thousands of years ago, that are bathed in suffering and blood, to seek meaning today?

Did Jesus Die on a Cross?
A brief glance at two of the world's great religions (Christianity and Islam) would tell us that Christianity teaches that Jesus died on the cross to save humankind from their sins while Islam teaches that this story is false. In the Christian New Testament we read, "For Christ also died for sins once and for all" (1 Peter 3:18). On the other hand, in Islamic Scriptures we read, "...they killed him not, nor crucified him, but so it was made to appear to them, and those who differ therein are full of doubts..." (Quran, sura 4, ayat 157-158).

Which religion is correct? Regarding this particular issue, one of these two religions must be based on false information related to Christ (and the divinity of Christ, or the lack of his divinity, is rather an important point). With more than a billion people as followers of each of these major religions, this is a huge number of people to be supporting false facts if we consider that this teaching (which is the very foundation of Christianity) must be wrong for one of these two religions. Or, could members of these two religions simply peacefully accept their different interpretations?

The Inspired Word of God?
Billions of people make important life decisions every day based on what they believe are the inspired words from God delivered to humankind at some point in the past (as written now in the Bible or in the Quran), and for most of these people, they rarely consider the origins of their beliefs. In the example of Jesus given above, the Holy Scriptures of Islam and Christianity cannot both claim to be the result of the inspired word of God because in this fundamental belief (the divinity of Jesus), they disagree.

A quick search on Google would tell us that there are literally thousands of religions spread throughout the world. Most of these religions have their own version (or interpretation) of Holy Scriptures which they believe were inspired by their God. Unfortunately, with apparent contradictions between various religions, either there is more than one God, or some religions have incorporated false facts into their beliefs.

While it might be easy to say, "to each to their own," the problem occurs when significant numbers of people are suffering throughout the world largely because of religious differences. Some countries are torn apart because of religions differences (think of Iraq and Syria in this context). Some people are persecuted and encounter various forms of prejudice because of their beliefs (think of just about any country in the world to support this thought).

The Importance of Our Beliefs

Our beliefs impact the way we live. Our beliefs contribute directly to our happiness. Our beliefs direct how we make decisions every day.

At a time when suicides in our world outnumber people killed in wars; at a time when anxiety and mental illness are on the rise; and at time when more and more people are rejecting traditional religions, what new beliefs will emerge in the 21st century to help people find happiness?

As a crisis counselor, I was once asked to see a young man who had attempted suicide by jumping off the roof of a two-story building. Let's call this young man Tim.

I visited Tim in a local hospital. Tim's attempt to kill himself left him with serious head and neck injuries, as well as a broken back.

Tim struggled emotionally and physically to tell me his story. Tim was from a conservative Christian family. At an early age he discovered that he was gay, but he didn't dare tell anyone because his religious family would have disowned him. Throughout his teenage years, Tim was plagued with guilt. He couldn't stop thinking about being with other guys, but a little voice inside his head (that had been shaped by a religion that was anti-gay) kept telling him that his thinking was sinful. His mind became a constant battlefield between what he desired and what his church had told him was right.

Unable to meet his own needs and plagued by the guilt of wanting something that was contrary to his beliefs, Tim became depressed, a depression that would eventually lead to him attempting suicide.

In the 21st century we know that our sexual orientation is not so much a choice, as it is a biological and genetic predetermination. Tim didn't decide to become gay. His genes determined his sexual orientation. Tim was born gay. For those who believe that God had some hand in creating our world, then it follows that God created Tim. If you believe that God created the world and everything in it, then God is therefore responsible for everyone's sexual orientation which is obviously a contradiction to the apparent word of God in the Old Testament of the Bible where being gay is condemned. Of course, there's another explanation that anyone's sexual orientation is the result of our genetic makeup, and God had nothing to do with this.

A Change of Beliefs to Find Happiness

For someone like Tim to find happiness, he is going to have to change his beliefs. If he wants to continue to believe in God, he will have to reject some teachings in the Bible concerning the sin of being gay. And this approach is not unusual in our world. More and more church attending people are beginning to realize that being a member of the LGBTQ community is a normal part of being a human, regardless of what the Bible (or Quran) says. And inappropriate beliefs from Holy Scriptures are not just limited to one's sexual orientation as we will explore further in the next chapter.

As you will see later in this book, some of the false teachings of various Holy Scriptures are being sanitized or even ignored in the 21st century. Some of the teachings from various inspired words of God are changing to better fit in with new learning in the 21st century.

As our understanding of the world around us changes, our belief systems may be challenged to change as well. The acceptance of the LGBTQ community is but one example of a false belief from the past changing in the 21st century to be more in line with scientific discoveries or new societal norms. For some people, the revelation that new facts may challenge past beliefs that originated from any Holy Scriptures will lead to abandoning the church of their parents and may even lead to rejecting the God of Holy Scriptures.

Finding meaning (and happiness) in the 21st century may necessitate examining, and in some cases shedding beliefs that are stuck in the past.

Viktor Frankl—a Holocaust survivor—wrote in his bestselling book *Man's Search for Meaning*:

> *"There is nothing in the world, I venture to say, that so effectively helps one to survive even the worst conditions as the knowledge that there is a meaning in one's life."*[27]

In this book we will explore some thoughts about how our past has shaped our present beliefs, but also some thoughts about how 21st century learning may reshape some of our beliefs as we seek to find meaning.

As our understanding of the world around us changes, are there new belief systems that would better help us to find meaning in the 21st century?

In the next chapter, we will begin to look at one of the number one why **questions–Why** do pain and suffering exist?

- 2 -
Why Do Some People Blame God for Pain and Suffering?

One Sunday night, I received a telephone call alerting me of a horrible tragedy, one where I would be required to provide counseling support. I was told that a teacher had just murdered his wife and baby, and then took his own life.

The next morning as I tried to help students who were in this teacher's classes, the expression of emotions which included rage, bewilderment, pity, sadness, confusion, disbelief, depression, anxiety, and even numbness was overwhelming. In a sense the school became paralyzed in grieving. While the students (and staff) had many questions, most of them began with the word why.

Barna, the public/opinion pollster was commissioned by author Lee Strobel to conduct a nationwide survey on the question, "If you could ask God only one question and you knew he would give you an answer, what would you ask?" The most common response was, "Why is there pain and suffering in the world?"[1]

Does God Test Us?

For many people, especially the religious, the answer to this question has something to do with God "correcting" or "punishing" us.

As a counselor I have worked with literally hundreds of unemployed people who were fired or lost their jobs through downsizing. Most of these people explained that losing their jobs was one of the most difficult things they had ever faced in their lives. Job loss can create a significant amount of grief. For some people the anguish can be as great as the death of a friend.

Louisa (not her real name) lost her job after working at a company for 26 years because the company went out of business. Louisa lived alone. She had divorced years ago and her two adult children now lived on their own.

I first met Louisa when her daughter brought her into the counseling office at the institution where I was working. Louisa's daughter explained that her mother lost her job the previous year. Since that time Louisa had virtually locked herself in her apartment, except for attending church several times during the week.

Louisa said that God was testing her. She said that God was trying to make her stronger by taking her job away from her. In her bargaining with God, Louisa had promised God that she would give more money to her church and improve her attendance if God would help her get a new job. While waiting for God to deliver on his end of the bargain, Louisa had made very few attempts to get a job.

Fortunately, with the support of her daughter and through counseling, Louisa learned that God had nothing to do with her job loss, and that giving money to her church and attending more was not going to help her get a new job. When Louisa eventually began to accept personal responsibility for what was happening in her life and took action to resolve her unemployment problem, she was able to find a new job.

Sometimes when people blame their problems on God, they fail to solve their problems. When people think that God is testing them by punishing them, they often fail to take personal responsibility for what is happening. Sometimes it's simply easier to blame God than it is to consider solutions that might involve taking personal action.

On a larger scale, after the terrorist attacks on the twin towers in New York City occurred in 2001, Laurie Goodstein, a journalist, reported the following in the *New York Times*, "The Rev. Jerry Falwell and Pat Robertson (both popular Christian evangelical leaders) set off a minor explosion of their own when they asserted on U.S. television that an angry God had allowed the terrorists to succeed in their deadly mission because the United States had become a nation of abortion, homosexuality, secular schools and courts, and the American Civil Liberties Union."[2]

This is a rather clear case of suggesting that pain and suffering are the result of God punishing people who disobey his teachings (or a clear case of religious leaders using a tragedy to promote their own agenda). Such erroneous thinking can be traced back thousands of years (and perhaps even hundreds of thousands).

Marni Jackson writes in her bestselling book *Pain—The Science and Culture of Why We Hurt*:

> *"The concept of pain as punishment turns up most vividly in the biblical story of Job, a wealthy, upright man whose faith in God is tested by Satan in a series of terrible afflictions. The idea of pain as spiritual punishment is still deeply entrenched in our attitude that physical pain arrives as a kind of test of moral character and should be toughed out."*[3]

I believe that there is a single reason, based on scientific fact, as to why most pain and suffering exist in our world and this reason has nothing to do with God correcting or punishing us. Before sharing this reason with you, let's look at an overview of why many people believe in the false idea that pain and suffering are caused by God.

After experiencing a tragedy, it's common to hear clichés such as: "God is testing me," "We all have a cross to bear," "It's God's will," "She's in a better place," "Even though you might not understand it now, God has a plan for you," "God needed another angel," "God never gives you more than you can handle," or perhaps the most often offered piece of advice, "Everything happens for a reason."

What if none of these clichés are actually true? What if pain and suffering have nothing to do with God?

Does Everything Happen for a Reason?

In 1755, a horrific earthquake in Lisbon killed more than 60,000 people. In the days following the earthquake, church leaders preached that the earthquake had been sent by God to punish the people of Lisbon for their sinning. As Voltaire argued, why would God punish Lisbon when there were other cities on the planet that were more deeply involved in sin? And one might also ask how such church leaders would explain the presence of earthquakes, hurricanes, tsunamis, and other natural phenomena that occurred on our planet long before humans arrived? Consider the asteroid that hit our planet around 65 million years ago wiping out a significant number of animals across the globe including the dinosaurs. Who was God punishing then? Had the dinosaurs been engaged in idol worship or other forms of sinning that enraged God?

God Sees the Little Sparrow Fall

As a child, I remember a song we used to sing in Sunday School that was probably one of the first experiences I can recall at church that I knew was not factually correct.

The song went something like this:

> *God sees the little sparrow fall.*
> *It meets his tender view.*
> *If God so loves the little bird,*
> *I know he loves me too.*

And then the chorus went like this:

> *He loves me too.*
> *He loves me too.*
> *I know he loves me too.*
> *If God so loves the little bird,*
> *I know he loves me too.*

As a child of perhaps 7 or 8, this little bit of religious indoctrination often filled me with confusion. It was obvious to me that the song writer had never seen a little sparrow fall out of a nest. I had. I had seen the splattered remains of little birds on the sidewalk as I walked to school. I had also seen the shredded remains of baby birds that had fallen to the grass instead of the pavement, where a dog or cat then proceeded to rip them to pieces.

As we sang that song Sunday after Sunday, my young mind told me that something was wrong with the words of the song. At this very early age, I most definitely understood that God had absolutely no interest in the welfare of any baby bird falling out of a nest. And if he didn't care about the little sparrows, did he really care about us?

Attributing pain and suffering to God has a lengthy history. Well-respected philosophers, writers and theologians have cemented this—what I believe to be false—concept. For example, the famous philosopher and theologian Thomas Aquinas (1225-1274) wrote that "God sometimes inflicts evil as punishment in order to maintain the just order of the universe."[4] Author C.S. Lewis wrote, "God whispers to us in our pleasures, speaks to us in our conscience, but shouts to us in our pain: it is His megaphone to rouse a deaf world."[5]

Some would point out that the source of such thinking is found in Holy Scriptures. For example, in the Bible we read, "For those whom the Lord loves he disciplines" (Hebrews 12:5-7), "My son, do not reject the discipline of the Lord" (Proverbs 3: 11-12), "Thus you are to know in your heart that the Lord your God was disciplining you just as a man disciplines his son" (Deuteronomy 8:5), and "Those whom I love, I rejoice and discipline" (Revelation 3:19).

What About Religions Other Than Christianity?

Upon further examination we would find that such thinking about pain and suffering is a part of other major world religions and not just exclusive to Christianity. For example, in the four Biblical quotes provided in the previous paragraph, the second and third quotes are from the Old Testament, so they also apply to the Jewish religion as well as Christianity.

Another major world religion is Islam. In terms of actual numbers, Christianity is the most popular world religion followed by Islam. Judaism is near the bottom of the scale in terms of the percentage of Jewish people compared to the overall world population, but because Christianity and Islam have their roots in Judaism, I have included thoughts on Judaism in this chapter.

According to the Pew Center for Research, Christians comprise 31.2% of the world's population; Muslims comprise 24.1% of the world's population; while the Jews comprise only 0.01% of the world's population. People with no religious affiliation account for 16% of the world's population[6]. According to research from the Pew Center, atheists and agnostics have almost doubled in the U.S. in the past 3 years. In considering this, keep in mind that atheists in

North America represent only 5% of the total number of atheists in the world (with China having the most atheists in the world).[7]

In Islamic Holy Scriptures we read, "We may test him" (Quran 76:3), and "We shall put you to test with some fear, and hunger, and with some loss of wealth, lives, and offspring" (Quran 2:155). Similar to Christianity, many Muslims also believe that pain and suffering can be attributed to God.

Christianity and Islam account for over half of the world's population. With such a significant percentage of the world's population being taught, through their scriptural teachings, that pain and suffering originate from God, it is no wonder that so many people have accepted this belief (we will look shortly at how this belief occurred long before the advent of Christianity or Islam). This impact is so pervasive that it's likely that many people who are not even members of Christianity or Islam have been influenced by such thinking.

The Gods Are Out to Get Us

The concept of linking God to pain and suffering has its roots in ancient pagan societies (and very likely back to the very beginnings of the human species in the distant past). The Romans believed in a multitude of gods and goddesses who had the power to influence everything from victory in war (Mars) to healing (Apollo). Poena was the Roman spirit of punishment and the attendant of Nemesis, the goddess of divine retribution, causing humans pain and suffering. The Greeks had a plethora of gods, although they had a few specific gods whose sole responsibility was to cause pain and suffering. These gods, known as the Algea, brought grief and sorrow to humans.[8]

Both the Romans and Greeks believed in gods who were active in their daily lives. These gods could bring good luck, or they could bring pain and suffering. The important point related to the ongoing discussion in this chapter is that these gods caused things to happen, that they were personally involved with humans. As a result, whenever bad things happened to people, there was an immediate perception that a god must somehow be involved.

If we were to continue back in time before the Greeks and Romans, we would find similar thinking in the religion of the Egyptians. Long before any of today's major religions appeared, the

Egyptians believed in gods that interacted in their daily lives in both good and bad ways. For example, Seth was considered to be the Egyptian god of pain and suffering, the god of chaos.[9]

One of the oldest religious traditions (besides the burial of the dead) is the act of making a sacrifice to please a God or gods. Throughout ancient historical times, sacrifices included everything from food offerings (which often included animal sacrifices, and in some cases even human sacrifices) to prayers and songs. And in the 21st century, for someone like Louisa whose story I shared earlier in this chapter, her sacrifice to try to please God was in the form of giving unrealistic amounts of money to her church.

Erik Hornung, a Professor of Egyptology at the University of Basel, wrote:

> *"Offerings celebrated deities' life-giving generosity and encouraged them to remain benevolent rather than vengeful."*[10]

The concept of sacrificing to a god in ancient times clearly implied that the gods interacted with people and that they had the power to either dispense punishments or rewards, a belief that still persists in the 21st century although we have replaced a pantheon of gods with just one God (or perhaps two if you give Satan equal billing, or three if you believe in the Trinity, or perhaps many more if you throw in the Virgin Mary or the Saints who some people pray to as though they were gods, and maybe even several hundred more when you consider that every modern day religion has their own unique version of God, and of course where do the angels fit into all of this?)

In looking at the religions of the Egyptians, the Greeks, and the Romans, we find a tradition of sacrificing to their gods. Sacrifices were considered an important part of pleasing their gods, of keeping the gods on their side, of restraining their gods from causing more pain and suffering. For example, the Egyptians tried to please their gods through animal sacrifices. Lucia Gahlin, an Egyptologist, wrote:

> *"Worship often involved making offerings to the gods*

*accompanied by invocation, in order to ensure
their continued and benign presence
in the lives of the people."*[11]

Robert Garland, Professor of the Classics at Colgate University, wrote in regards to the Greeks:

*"Sacrificial victims were then slaughtered
to the gods in the hope of securing their goodwill.
The Spartans, for instance, drove whole herds of goats
onto the battlefield for sacrifice."*[12]

The Romans believed in a concept known as "do ut des" (I give that you might give) which represents the reciprocity of exchange between the gods and humans. In other words, the Romans believed that by sacrificing to their gods that their gods would be pleased and bring them good luck. Jorg Rupke, a German scholar of comparative religion and classical philosophy, wrote this about the Romans:

*"The gifts offered by the human being take the form
of sacrifice, with the expectation that the god will
return something of value, prompting gratitude
and further sacrifices in a perpetuating cycle."*[13]

If we were to travel back even further in history, we would find that humans who lived many years before the Romans, Greeks, and Egyptians also practiced various forms of sacrifice to please their gods. While archaeological timelines are constantly shifting backwards as new discoveries are made (and new scientific methods for analyzing the finds improve), the burial of the dead accompanied by the presence of relics in gravesites occurred at least 30,000 years ago and likely much further.[14]

Red ochre, which might have been part of a funeral ritual, appears on some skeletal remains dating back approximately 250,000 years ago in the finds at Qafzeh, Israel.[15] Ancient burial practices that include tools or carvings suggest a belief in an afterlife which again supports the concept that humans believed in a personal relationship with their gods.

The concept that pagan gods caused things to happen provided a philosophical base for today's major religions to build on. This is a concept that has no factual basis, but nonetheless it is deeply rooted in the minds of most humans, even those who are not part of any organized religion. It is a concept that connects pain and suffering to a god or God, and it is also a concept that teaches that humans can have a personal relationship with their gods or God.

Does God Reward Or Punish Us?

Throughout history the concept that gods or God punished wrongdoings (or rewarded us when we obeyed) became a great way for religious leaders to control their members. It was a lot easier for the Egyptian Pharaohs to get their subjects to build massive pyramids or temples when the people were taught that the gods would punish them if they didn't obey (especially considering that the Pharaohs were viewed as being gods and could unleash immediate punishment).

It was a lot easier for the leaders of the Catholic Church, as an example, to teach their followers that they needed to build huge cathedrals to honor God, while it was the church leaders who became richer and richer while the common person continued to live in ignorance and poverty.

Throughout history church leaders often maintained the fundamental purity of their particular religion by convincing their members that God was always watching, and that he would punish or reward people based on them following the rules and doctrines of the church. And when churches have beliefs that relate to the doctrines of heaven (eternal bliss) and hell (eternal damnation), these can be extremely powerful concepts that can be used to control their members.

For thousands and thousands of years, humans have believed that we can speak on a personal basis to our gods. Most people have also believed for thousands of years that our gods, like a wise parent or good teacher, employ both punishments and rewards to shape our lives. This thinking is ingrained in us, whether we attend church or not.

Such thinking, as part of religion, is often simply accepted and rarely challenged. Darrel W. Ray, a psychologist, writes:

> *"For thousands of years, religion penetrated societies, largely unexplained and unchallenged. It simply existed. Those who attempted to question or expose religion were often persecuted, books burned, excommunicated, or even executed."*[16]

If the concept that gods or God cause pain and suffering has been with us for thousands of years, how does this impact us today?

God's Will and Plan for Us

There are a number of concerns related to the belief that God is in control of our lives (and that he uses punishments and rewards to shape us). First of all, by believing that God is controlling what is happening in our lives, we might fail to take responsibility for our thoughts and actions because we believe that whatever happens in our lives is the fulfillment of God's will and plan for us (and pain and suffering are his attempt to keep us on track).

Secondly, if we believe in God, there is a tendency for us to also accept the presence of the devil or Satan. And keep in mind that if God is the creator of everything, then he must have created the devil or Satan. Can you imagine some loving parent in the 21st century going to a prison to find the most vicious, cruel, criminal imprisoned there to be their child's godparent so that this criminal could test and punish their children to make them better? A ridiculous thought, isn't it? Yet, for those who believe that God created Satan to test them, this is exactly the same kind of thinking. Believing in Satan results in a double whammy in the area of pain and suffering. First we have God using pain and suffering to discipline us, then thrown into this mix is the belief that God also allows Satan to cause pain and suffering. For example, in the Bible we read, "Your enemy the devil prowls around like a roaring lion looking for someone to devour" (1 Peter 5: 8-9). In the Quran we read, "Anyone who follows the steps of Satan should know that he advocates evil and vice" (Quran 24:21).

When people believe in God and Satan, it's far easier for religious leaders and even politicians to convince their followers that wars, persecution, and ethnic cleansing are righteous. Wars become Holy Wars, an attempt to defeat the Great Satan who is leading the other side.

Let's look a little further at how the belief that God and Satan causes pain and suffering can affect us. Eleanor Roosevelt, American First Lady from 1933-1945, suffragist and activist for civil rights, stated:

> *"In the long run, we shape our lives, and we shape ourselves. The process never ends until we die. And the choices we make are ultimately our own responsibility."*[17]

Have you ever heard anyone say, "It was God's will," or maybe even, "The Devil made me do it," after they made a poor decision or things didn't go the way they were hoping. Blaming God (or even Satan) when things go wrong may provide what appears to be the perfect scapegoat, but such beliefs can prevent a person from ever taking ownership of their own problems.

In a counseling relationship I worked with a woman whose second husband had just left her (let's call her Carol). When I met Carol, she was in her late twenties She wanted help because she was depressed and had suicidal thoughts. During our first few sessions, whenever I asked Carol to tell me a little more about the problems she had encountered in her marriages, her response was always, "What happened was God's will."

Eventually, when this reason was explored further, Carol explained that she had stopped going to church when she was around 14 or 15, and since that time God had punished her with an assortment of problems, including two failed marriages, in his attempt to get her to return back to her church.

Such thinking on Carol's part had many flaws. Further discussion clearly showed that her problems had nothing to do with God punishing her, or God attempting to coerce her to return back to a church.

Carol married her first husband after knowing him for only one month when she was 21. They had little in common, other than they enjoyed sex together. She explained that their marriage was a constant battle. She and her husband couldn't agree on anything. Intense arguments soon replaced their sex. Carol's tumultuous marriage to her first husband lasted a little over two years before they separated and subsequently divorced.

A few months after her divorce, Carol married again. Her second marriage lasted almost three years before her husband left her for another woman. Once again, she blamed the marriage breakdown on God. In her second marriage, Carol's husband was an abusive alcoholic, a man who beat her and crushed whatever self-esteem remained from her failed first marriage. When Carol was asked why she stayed with an abusive husband for such a long period of time, she replied, "God was testing me. He wanted me to grow stronger. He wanted me to return back to church." With such thinking, Carol allowed herself to be punched, to be insulted, and to suffer constant emotional pain.

After her second husband left, Carol headed back to church. She decided that she would get right with God before finding another husband. For almost a year, Carol lived a celibate life, a life of prayer and a life where she found refuge in participating in every possible church activity that was available to her. It was during this time that Carol began to suffer depression and developed suicidal thoughts. She said that the more she prayed, the worse she felt. Guilt plagued her. She believed that she had let God down because she had failed in her marriages.

Although Carol initially blamed God for her marriage problems and her unhappiness, through counseling she eventually began to see that her problems were the result of making inappropriate decisions, rather than being the result of God punishing her. She also began to understand that the depression she experienced was directly influenced by the guilt she suffered in not being able to live up to the expectations of what she thought God expected of her.

Once Carol was able to remove God as the cause of her pain and suffering, she was able to achieve more emotional stability and improve her self-esteem through her work and improved relationships with others.

Such a process does not happen overnight, and it does not happen without a commitment to change (and often it does not happen without professional help). As outlined throughout this chapter, the belief that God punishes or rewards us is deeply ingrained in our thinking. To eliminate such thinking and to take personal responsibility for our successes and failures can take a concerted effort.

God Helps Those Who Help Themselves

There is an old saying that "God helps those who help themselves" (similar thoughts having their roots in ancient Greek tragedies—Sophocles in *Philoctetes* as an example in 409 BCE—and made more popular by Benjamin Franklin in his Poor Richard's Almanack in 1736). If you accept such a statement, upon further thought you might realize that this statement is really saying that God doesn't help at all.

Finding happiness and success (and meaning) begins when we take responsibility for what is happening to us, when we realize we are in control of our destiny. Our lives are not the fulfillment of God's will, they are the fulfillment of our choices.

Our Thoughts and Prayers

Let's look at another example to illustrate what we have been exploring. On November 5th, 2017, a lone gunman murdered 26 innocent people in a church in the small rural community of Sutherland, Texas. In listening to the initial responses on the news, I heard several people say, "Our thoughts and prayers are with the families of the victims."

One of the most disturbing statements uttered came from a woman who was talking to CNN's Chris Cuomo. She said, "The devil never rests."[18] Instead of placing the full blame on the shooter (and perhaps ineffective gun laws), Satan was somehow involved in the shooting and through thoughts and prayers somehow everything would be okay again.

The day after the Sutherland tragedy, on a broadcast of *CNN Tonight*, host Don Lemon opened his show with a monologue arguing that 'thoughts and prayers' were not enough of a response to the previous day's horrendous shooting in Sutherland, Texas.[19] When people believe that prayer will provide an answer or that such tragedies were somehow part of God's will or were caused by the devil, then there are never going to be solutions to prevent such gruesome killings from happening again.

Hiroshima and Nagasaki

For anyone who is still not convinced that God has anything to do with such tragedies, then perhaps they might try to explain how it was God's will for atomic bombs to fall on Hiroshima or Nagasaki in 1945. How could these tragedies be explained as a punishment from God? How could the deaths of more than 129,000 innocent people (and the incredible suffering of the wounded who were often severely burned)[20] in these disasters be blamed on God?

And for anyone who believes that God is all-powerful, the next question might be, "Why doesn't he intervene to stop such horrendous tragedies?" If your answer to this question is that God doesn't intervene because he wants us to have free will, then such an answer is a contradiction to the belief that God is in control of what happens in our world. If God is in control, then you can't have free will. How could you possibly fight against the force that created the universe? And if you believe that you have free will, then you cannot believe that God is in control of all that happens.

In looking at the bombing of Hiroshima and Nagasaki, perhaps the cause didn't have anything to do with God punishing anyone. This tragedy was rather the result of religious and political leaders (on both sides of the war) convincing their followers that they needed to defeat an opposing evil power. In wars, persecution, and ethnic cleansing, the real cause is often related to one side thinking that their beliefs are more righteous than those of the other side, and the other evil power has to be defeated.

The Great Satan

The Islamic Jihadist believes that the U.S. is the Great Satan, an evil power that needs to be defeated (a term first used by Iranian leader Ruhollah Khomeini in a speech on Nov. 5, 1979).[21] Sometimes the evil that needs to be confronted is in the form of a political ideology. For example, since the end of the Second World War the U.S. has been engaged in a fight against godless communism which represents the Great Satan.

After the Second World War, popular Evangelical Protestant leader, the late Reverend Billy Graham often aimed his sermons against Communism. Pulitzer Prize winner Frances Fitzgerald quoted Graham as saying, "Communism is a religion that is inspired, directed and motivated by the Devil himself who has declared war against Almighty God."[22] Such thinking by Americans eventually contributed to the U.S. involvement in Vietnam to stop the spread of evil communism, a war that led to approximately 1,353,000 civilian and military deaths, not to mention the incredible pain and suffering that ensued for the survivors.[23]

Although the U.S. failed in its attempts to stop the spread of communism in Vietnam, today Communist Vietnam and Christian U.S. are good friends. When the U.S. discovered that they could benefit financially by trading with Vietnam, U.S. leaders quickly forgot about the evil godlessness of communism.

The Persecution of Christians

Yuval Noah Harari, a professor in the Department of History at the Hebrew University of Jerusalem, in his International bestselling book *Sapiens: A Brief History of Humankind,* talks about the Roman persecution of early Christians and then goes on to say:

> *"Still, if we combine all the victims of all these persecutions, it turns out that in these three centuries, the polytheistic Romans killed no more than a few thousand Christians. In contrast, over the course of the next 1,500 years, Christians slaughtered Christians by the millions to defend slightly different interpretations of the religion of love and compassion."*[24]

Often, it's not God who is causing pain and suffering, but rather religious leaders using God to rally their followers to fight against those who interpretation of God or Holy Scriptures might be slightly different.

On August 23, 1572, French Catholics attacked French Protestants (an attack known as the St. Bartholomew's Day Massacre). Between 10,000 and 70,000 Protestants were eventually murdered by the Catholics. Upon hearing the news of the slaughter, the Pope in Rome was so overjoyed that he organized festive prayers to celebrate the occasion and commissioned Giorgio Vasari to paint a fresco on one of walls in a room in the Vatican to illustrate the massacre (a room that is currently unavailable to visitors).[25] This is but one example of Catholics and Protestants killing each other, causing enormous pain and suffering in the name of God. With both sides believing in the same God, how could God possibly be responsible for the horrendous suffering that occurred?

The persecution of Christians (often by other Christians), is another way that God became associated with pain and suffering.

The Persecution of Jews

During the Second World War, more than six million Jews were slaughtered in Europe because of their religious beliefs. Ever since the war, much of the world has remained in shock that Hitler and the Nazis could be so inhumane, yet the persecution of the Jews had its roots in Christian theology dating back hundreds of years before the Holocaust occurred. The Jews, often blamed for the killing of Jesus, have been persecuted by both Catholics and Protestants for a significant period of time throughout history.[26] What the Nazis did to the Jews was a continuation of the horrors inflicted on Jews by Christians, all in the name of God.

Although the Nazis required the Jews to wear a yellow Star of David during the war so that they could be instantly recognized (it's a lot easier to persecute a group of people when you make them wear something that makes them stand out), this practice wasn't invented by the Nazis; it was first used by Christians during the Inquisition hundreds of years ago. In the Synod of Narbonne in 1227, we read, "That Jews may be distinguished from others, we decree and empathetically command that in the center of the breast (of their garments) they shall wear an oval badge,"[27] (and even earlier than this, the Muslims in the early 8th century required Jews to wear two yellow badges).[28]

On July 15, 1205, Pope Innocent III stated that Jews were doomed to perpetual servitude and subjugation because they had crucified Jesus.[29] In 1234, Pope Gregory IX continued Innocent's work through the Decretals which further elaborated on perpetua servitus iudaeorum—perpetual servitude of the Jews—with the full backing of the church. Gregory condemned the Jewish Talmud as blasphemous and heretical. He also instituted the Papal Inquisition, an arm of the church that would eventually severely punish heresy.[30]

If Jews didn't renounce their religion and accept Christianity they could be persecuted. Gerald S. Sloyan, professor emeritus of religion at Temple University writes in an article on the U.S. Holocaust Memorial Museum website:

> *"In the years 500-1500 the Jews, as a religious and cultural minority, were often preyed upon by the Christian majority."[31]*

Through the persecution of Jews for hundreds of years (often by Christians and Muslims), God once again became associated with pain and suffering.

Catholics and the Inquisition

The Medieval Inquisition, which began in France in the 12th century and continued into the mid-15th century, although some forms of the Inquisition existed into the early 19th century (and in 1908 the Inquisition was given the new name of "Supreme Sacred Congregation of the Holy Office" by the Catholic Church, and in 1965 it became the "Congregation for the Doctrine of the Faith")[32] was an attempt by the Catholic Church to combat those who didn't believe in the doctrines of the church (which included both Jews and other non-believers).

Punishments against non-believers during the Inquisition included the horrendous burning at the stake, life imprisonment, banishment from a community, and torture (which included starvation, placing burning coals on body parts, hanging by the arms, being stretched on a rack, mutilation, sitting on wedges covered in spikes, the crushing of body parts, and ripping away flesh).[33]

The Inquisition which lasted for hundreds of years and which caused tremendous fear in the name of God helped to cement the belief that pain and suffering were associated with God, and probably terrified many generations into converting to Christianity.

In the book *God's Jury: The Inquisition and the Making of the Modern World*, the author Cullen Murphy writes:

> *"The Inquisitors accepted without doubt that God paid close attention to the affairs of human beings*

and was active in the Inquisitor's cause."[34]

And make no mistake about it, if the Inquisitors believed that God was supportive of the pain and suffering they inflicted on innocent people then the people being tortured (or who watched others being tortured) also believed that pain and suffering were associated with God. To make such punishments appear to be sanctioned by God, Pope Alexander IV in 1252 decreed that inquisitors could clear each other from any wrongdoing that they might have done during torture sessions.[35]

Protestants and the Continuation of Persecution in the Name of God
While the Catholic Church may have been the driving force in the Inquisition and the Crusades, the Protestant Church is not blameless when it comes to causing pain and suffering in the name of God. For example, John Calvin—an influential leader in the early Protestant Church—was instrumental in the death of Michael Servetus in 1553 who was burned alive because he disagreed with some of Calvin's interpretation of biblical doctrines.[36]

Concerning violence by Protestants in the name of God, D.M Bennett in 1878, wrote:

> *"Protestants have pursued Catholics: Lutherans have hunted Anabaptists: Episcopalians have burned Puritans; Puritans have hanged Quakers; Calvinists have tortured Unitarians; and all have united in persecuting the heroic Infidels who have refused to believe in any of the multifarious and conflicting creeds."*[37]

Even Martin Luther, the great founder of Protestantism, said: "I, Martin Luther, slew all the peasants in the rebellion, for I said that they should be slain; all their blood is upon by head. But I cast it on the Lord God, who commanded me to speak this way."[38]

Once again, we see that pain and suffering are linked to God and religion.

Luther also "denounced the Jews in a series of pamphlets written in vituperative fury."[39] At the Diet of Worms (diet being the word for an assembly meeting and Worms is a city in Germany), Luther said that all Jews should be driven from Germany. He even authored a book titled *On the Jews and their Lies*. Luther described the Jews as a 'brood of vipers', the exact same phrase used many years later in a speech when Hitler said:

> *"My feeling as a Christian points me to my Lord and Savior as a fighter. It points me to the man who once in loneliness, surrounded by a few followers, recognized these Jews for what they were and summoned men to fight against them. In boundless love as a Christian and as a man I read through the passage which tells us the scourge to drive out of the Temple the brood of vipers and adders. How terrific was His fight against the Jewish poison... and if there is anything which could demonstrate that we are acting rightly it is the distress that daily grows. For as a Christian I have also a duty to my own people."*[40]

Religion, God, and War

Lest we think that the association of war and religion is something from the past, consider these words from Rev. Dr. Mae Elise Cannon in a 2014 *Huffington Post* article:

> *"A couple of weeks ago I was at an interfaith gathering wrestling with questions about religion and peace. Someone, an elderly faith-filled individual who has worked in inter-religious dialogue for decades, suggested that globally we are moving into a time where religious persecution and oppression is increasing around the world and toward numerous religious communities. Religious fanaticism seems to be ruling the day."*[41]

Former British Prime Minister Tony Blair was quoted as saying "Extremist religion is at root of 21st century wars."[42]

While Christians in the western world often see extremist Islam as the problem, these same people blithely continue to sing songs like *Onward Christian Soldiers* with the words:

> *Onward Christian Soldiers!*
> *Marching as to war,*
> *With the cross of Jesus*
> *Going on before.*
> *Christ, the royal Master,*
> *Leads against the foe;*
> *Forward into battle,*
> *See his banners go.*

Although some Christians who sing *Onward Christian Soldiers* might argue that they are singing about their battle with Satan, it would seem to be more apparent that what they are really trying to do is fight against anyone who doesn't hold the same beliefs as them.

Having been a member of an evangelical church, for the first 30 years of my life, there was little doubt in my mind that when we sang such a song it was intended to be a battle cry to save anyone who didn't believe in Jesus. When one religion suggests it is somehow superior to other religions, only bad things can happen. As a result, this can provide more fuel causing people to believe that God is connected to pain and suffering.

Does God Send Diseases?

The idea that God (or gods, and the devil) is responsible for causing pain and suffering goes beyond war and persecution. Disease is also often blamed on God. For example, a study reported in the *Huffington Post* stated:

> *"People were more likely to believe that God gave*
> *them cancer than admit their poor diets and sedentary*
> *lifestyles may have played a predominant role."*[43]

Such thinking is not new. In 1918, a flu pandemic infected more than 500 million people around the world resulting in the deaths of 50 to 100 million people. This was one of the greatest disasters in human history associated with disease. Some people attributed this disaster to God. From *"Historically Speaking"*, Howard Phillips writes:

> *"Even more revealing for the historian is that such explanation also sought to account for why God had sent the epidemic. Punishment for sin was the most common reason offered."*[44]

The belief that God causes diseases has its roots in ancient civilizations. As explained earlier in this chapter, for many thousand years humans sacrificed to their gods because they believed that this act would please their gods and keep the gods from inflicting pain and suffering on them.

Earthquakes, Hurricanes, and Tsunamis

Many humans have also had a belief instilled in them that God (or one of their gods) causes natural occurring phenomenon. For example, the Haida Native Americans of the north Pacific Coast believe that earthquakes are caused by movement of their deity's large dog who lives under the land.[45] Hawaiian mythology records that the eruption of volcanoes are caused by the anger of the goddess Pele.[46] The Greeks believed that Poseidon, when he was angry, caused earthquakes.[47] While most modern humans are not part of the Native Americans, or they don't believe in Hawaiian mythology, beliefs linking God to natural disasters are still ingrained in the thinking of many in modern societies.

In more recent times a tsunami struck Indonesia in 2004 killing between 230,000 to 280,000 people. In spite of the scientific facts explaining the cause of such a disaster, there were still some who saw this tragedy as an act of God. Dr. Erwin W. Lutzer, evangelical pastor and author, wrote, in reference to the tsunami, "God is the ultimate cause of all natural disasters. The Bible explicitly traces natural disasters to the hand of God."[48]

Believing that God is the ultimate cause of natural disasters has serious implications for those of us living in the 21st century. We are living at a time when rampant climate change could impact the future of humanity and much of life as we know it (more on this in Chapter 9). In considering the possible devastation that could result from climate change, we find that those who continue to believe that God punishes and corrects us, and that God is responsible for natural disasters that might occur throughout the world (including any that result from climate change) might lack the desire to fight such problems.

As has already been explained in this chapter, believing that God sends us pain and suffering takes away any sense of personal responsibility to prevent such discomfort, and for some this can even prevent taking positive action to deal with the pain and suffering.

From the earliest attempts of our ancient ancestors to understand the world around them, throughout history after the era of Christ, and until today, the concept that God is directly connected to pain and suffering has been ingrained in the thinking of humans.

In the next chapter we will look at the actual causes of pain and suffering, scientific based facts that have nothing to do with God or the devil (or religion).

- 3 -
What Does 21st Century Learning Teach Us About Pain and Suffering?

In the previous chapter we looked at why many people believe that pain and suffering are the result of God punishing or correcting us. In this chapter, we will look at how 21st century learning explains pain and suffering. Similar to a movie with a warning of violence at the beginning of it, I would like to caution readers that some of the content of this chapter is graphic and can be disturbing. I would also like to note that for those who find this chapter bleak or depressing, that before this book ends I will attempt to renew your hope. After all, this book is all about our search for meaning in the 21st century and I most definitely intend to leave you with some positive possibilities.

I live in a neighborhood where it's common to see rabbits running around. For the most part, the rabbits are cute to watch although they can also be annoying when they eat some of the emerging spring flowers or strip the bark off newly planted trees. Over time, I've learned which spring flowers they avoid (daffodils) and how to save new trees (by wrapping white spiral tree guards around the trunks).

Rabbits are the one animal (perhaps after puppies and kittens) that have a high cuteness factor. It's rare to find someone who doesn't smile upon seeing a loveable bunny. In fact, the cuteness factor of rabbits is so high that endless books and movies have featured them. For example, Goodreads currently lists 151 of the best storybooks about rabbits. Disney often capitalizes on the popularity of rabbits with them having a prominent role in movies such as Bambi which featured Thumper, or Roger Rabbit, and even the White Rabbit in Alice in Wonderland. Rabbits such as Bugs Bunny have starred in cartoons, and we even have the Easter bunny.

When most people see a baby rabbit they are generally reminded of the animated characters they see in films or read about in books, even if this is at a subconscious level. As we look at adorable bunnies it's almost as if we expect them to smile or start talking to us. Other than dogs or cats, rabbits have probably been anthropomorphized more than any other animal. We have been so conditioned to see their cuteness that we often fail to notice anything else about them.

Over the years, we've had a number of rabbit nests in our backyard. My first close-up experience with baby rabbits occurred a few years ago when I noticed a red-tailed hawk perched on our back fence several days in a row. Then, one morning as I was eating breakfast, the hawk dove into our yard and flew away with a tiny creature struggling to free itself from the hawk's sharp claws. Upon further inspection, I found a nest of seven little bunnies under a clump of dead grass where the hawk had snatched one of them. They had big bulging eyes just like the rabbits in cartoons. Instant cuteness.

Although I didn't want to touch the baby rabbits (the smell of a human on their skin could prevent their mother from returning to care for them), I was curious to look a little closer. As I knelt down, I immediately saw red dots moving over the skin of the baby rabbits. The bunnies were infested with tiny red mites—thousands of them.

After doing a little research, I discovered that there are typically three different types of mites that live on rabbits in the wild: ear mites, fur mites, and burrowing mites. As their names suggest, some of these mites live on the surface skin of the rabbits while others actually burrow into the skin. Burrowing mites dig into the skin and make small tunnels where they lay their eggs.

These parasites can also infect dogs, cats, and even humans, and are present on the skin of most mammals in the wild. Each parasite can suck up to three times its body size from its host animal. When animals scratch at the mites, this can lead to self-mutilation, wounds and secondary bacterial infections.

Mites are among the most diverse and successful creatures on our planet. It is estimated that there are some 48,200 species of mites that feed on animals and plants.[1, 2]

21st CENTURY LEARNING ABOUT PAIN & SUFFERING / 59

Bacteria and Our Pets

For some readers, you might already be wincing at the thought of tiny red mites crawling all over the skin of a rabbit, but, as the infomercials often say, "Hold on. There's more."

Mites are not the only small creature that can infest the skin of animals. Most animals are crawling with bacteria, both on their skin, inside their mouths, and inside the gut. In a research study at North Carolina State University, the researchers looked at the homes of 40 families. They found that families with dogs had 42 percent more bacteria groups living on pillowcases in the home than non-dog homes.[3] Consider cats who use their paws to bury fecal matter in their litter box and then proceed to walk around your house (and possibly even on your kitchen counter) spreading germs from their paws onto anything they touch (including you).[4] Diseases such as Salmonellosis can be traced back to cat litter boxes. In addition, cat excrement is the home to a horde of parasites such as hookworms, roundworms, and ringworms that can also infect humans.[5]

Fortunately, there's a possible benefit to having pets in your house. Consider the following, "If the latest research on pets and human health is correct, that cloud of dog-borne microbes may be working to keeping us healthy. Epidemiological studies show that children who grow up in households with dogs have a lower risk for developing autoimmune illnesses like asthma and allergies, and it may be the result of the diversity of microbes that these animals bring inside our homes."[6]

But, make no doubt about it: diseases can and do come from harmful bacteria invading our bodies. Sickness, pain, and suffering, while in some cases having a genetic cause, are also the result of harmful bacteria and viruses, and in some cases random accidents. Pain and suffering can generally be linked to a specific cause, and punishment from God is not one of the causes. Our bodies are in a constant fight for survival in the same manner as the tiny bunnies I saw who were infested with mites.

Colonies of Bacteria Living In and On Us

It is not only our pets who are covered with bacteria. As humans, we are also colonized by many microorganisms that live on our skin and inside our bodies. It is estimated that the average human body has more non-human cells than human cells. A press invitation for a conference on *The human being - a bacteria controlled superorganism* begins:

> *"Rather than just being a single organism, the human body is host to trillions of bacterial organisms that interact and influence our development and function from the moment we are born."*[7]

Some researchers have estimated that we have ten times more non-human cells than human cells,[8] and it's estimated that we each have in excess of 37 trillion human cells.[9] It is estimated that 500 to 1,000 different species of bacteria live in our gut.[10] In addition, each of us sheds approximately 39 million bacteria into our environment every hour (time to get out the vacuum cleaner).[11]

Of course, it should be noted that for the most part these microbes (such as bacteria and yeasts) help to keep us healthy (although poor hygiene and injuries can lead to infections and other diseases from these bacteria). Even though our skin and insides are the home to trillions of non-human cells, under most conditions, these invaders help us, rather than hurt us. These foreign cells actually protect us as they fight against infections that could cause an illness. In addition to this daily battle occurring in every human, our white blood cells are engaged in a war against infectious disease and foreign invaders. We also have natural killer cells within us that destroy any cells that have been infected by viruses or tumors.

The Daily Battle to Survive

Every creature on our planet is in a constant battle for survival. In the book *The Call of the Wild*, the author Jack London writes:

> *"He was a killer, a thing that preyed, lived, unaided, alone, by virtue of his own strength and prowess, surviving triumphantly in a hostile environment where only the strong survive."*[12]

Let's return back to rabbits in the wild. Even the cutest bunny is in a fight for survival every day. Rabbits are constantly under attack, sometimes from dogs or even humans, but also from diseases caused by the infestation of bugs and various forms of bacteria. They are also threatened by environment factors such as pollution, pesticides, and changes to their habitat.

If we are to consider the real cause of pain and suffering in our world, let's first begin by looking at animals (other than humans for the moment) because they are free from the restraints and biases of religion. While many humans might believe that God sends pain their way as a punishment or a form of correction, it's unlikely that these same people believe that God sends pain and suffering to animals in the wild because he is punishing them. All animals, from single-celled organisms to massive animals such as the blue whale are in a constant, daily battle for survival. It follows logically that the same natural laws that apply to the survival of all other animals on our planet must also apply to us. It doesn't make any sense to say that natural causes can result in pain and suffering throughout nature, but somehow it is God who causes pain and suffering in humans.

Have you ever heard people say things like, "Why can't humans get along like animals in the wild?" or "Mother Nature is perfect," or even "Why can't we learn from nature?" Unfortunately, people who say such things have rarely taken a closer look at what really happens in the wild. Perhaps, their view of Mother Nature has been falsely shaped by Disney animations. Animals in the wild are in a constant fight for survival. Let's explore this further.

Cannibals Around Us

I received a small aquarium when I was ten. My first fish were guppies. Guppies are a fascinating animal to observe. Male guppies arch their bodies in a courtship dance before they mate with a female. Once pregnant, you can watch the development of the embryos to some degree because the belly of a female guppy is transparent (the gestation period for guppies is generally 3 to 4 weeks). As the female guppy's belly swells in size, it doesn't take very long before you can actually see the eyes of the tiny babies through the translucent skin of the mother. When the day comes for the guppies to be born, they drop from the mother one by one (with typically between 30 to 60 babies over a few hours), sometimes a few seconds in between each birth, sometimes a few minutes.

Baby guppies immediately begin to wiggle in their first attempts to swim after leaving their mother. I remember watching my first live birth. I was excited to see the miniscule babies falling from their mother. My excitement quickly changed to horror. Suddenly the other guppies in the aquarium attacked the babies. Within a few minutes, most of the baby guppies had been eaten by the other guppies before I was able to rescue a few of the babies with a spoon and place them in a jar of water.

Guppies are not alone in cannibalizing their young. There are recorded instances of more than 1,500 species of animals cannibalizing their young.[13] Cannibalism can be the result of hunger, overcrowding and even sexual conflict where one animal kills the offspring of another animal to prevent the spread of this animal's genes.[14]

Not only do some species of animals cannibalize their young, sometimes embryos fight for survival while they are still in the womb. For example, sand shark embryos begin to cannibalize each other in utero about five months into their year-long gestation, which often results in only one baby shark eventually being born.[15]

As well, many animals eat the young of other animals. You might recall that earlier in this chapter, I talked about a red-tailed hawk swooping down on a baby rabbit in a nest in our backyard.

First Flight

Another animal that frequents our yard is the mourning dove. We have several hanging planters with flowers in our yard in the summer. Mourning doves love to build nests in these planters. Because the planters are at about eye level, we are able to watch the development of the babies after they hatch.

It is particularly interesting to watch the baby doves as they learn to fly. Often the baby doves (there are generally one or two of them) can be seen sitting on the edge of one of the hanging planters, beside one of their parents before they launch themselves.

On one occasion I was watching two baby doves sitting in a hanging planter with their parent. The baby doves sat there flapping their wings as though they were getting ready to fly. Finally, one of the plump babies jumped from the planter and wildly flapped its wings. About ten feet into its solo flight, suddenly there was a burst of feathers. You guessed it: that red-tailed hawk was back again. As a clump of feathers floated to the ground, the hawk flew away with the baby dove secure in its claws.

Violence in Nature
While there is a tendency among some people to believe that only humans can be cruel to each other, consider some of the following examples of violence in nature.

Let's start with the praying mantis. As a male copulates with a female, the female may turn and bite its head off. Ants in their wars with each other are known to tear each other limb from limb (and they also take other ants as slaves and then eat them once they are finished with them). Ichneumon wasps paralyze their victim and then lay an egg inside it, with the larva eventually eating the host animal from the inside out. If a male lion wants to have sex with a female who resists because she is caring for her cubs, the male lion will kill the cubs in order that he can have sex with the female. At the Toronto zoo in 2011, Aurora—a popular female polar bear—killed two of her adorable cubs (and the previous year she killed some cubs at birth by partially eating them).[16]

Cuckoo birds lay their eggs in the nests of other birds. When the eggs hatch, the baby cuckoo birds push the other baby birds out of the nest to their death. Even male dolphins, those wonderful loveable creatures that we pay exorbitant amounts of money to swim with, have been known to kill their young in order that the female dolphins are more available for sex. Chimpanzees have been filmed in the wild snatching babies from their mother's arms and then eating them. Household cats have been known to kill birds and small rodents, but don't always eat them. Jeff McMahan, professor of philosophy at Rutgers University, writes:

> *"Viewed from a distance, the natural world often presents a vista of sublime, majestic placidity. Yet beneath the foliage and hidden from the distant eye, a vast, unceasing slaughter rages. Wherever there is animal life, predators are stalking, chasing, capturing, killing, and devouring their*

prey. Agonized suffering and violent death are ubiquitous and continuous."17

Survival of the Fittest

What happens in the natural world around us is evidence of the survival of the fittest (a phrase that originated from Darwin's theory of evolution). This is a basic law that applies to all living organisms. The phrase describes the mechanism of natural selection. Natural selection, as defined in the American Heritage Dictionary, is:

> *"The process in nature by which, according to Darwin's theory of evolution, organisms that are better adapted to their environment tend to survive longer and transmit more of their genetic characteristics to succeeding generations than those that are less well adapted."18*

Pain and suffering in the world are largely the result of the law of natural selection (along with natural disasters such as hurricanes which are also based on scientific principles rather than being a punishment from God).

Animals encounter pain and suffering, not as the result of God punishing them but as a result of every creature's fight for survival, the innate desire to propagate their gene pool. It is this same instinct in humans that causes pain and suffering. Pain and suffering are not the result of God punishing us (or other animals); they are the result of being part of a living world where every creature is in a constant battle for survival. As bestselling author Haruki Murakami writes:

> *"Genes don't think about what constitutes good or evil. They don't care whether we are happy or unhappy. We're just means to an end for them. The only thing they think about is what is most efficient for them."19*

Although it is not my intent to provide a detailed overview of natural selection or evolution, let's at least consider a few examples that might help any readers who have little background in this area. Readers should also note that I believe that evolution is a scientific fact, perfectly stated by Richard Dawkins, an award-winning ethnologist and evolutionary biologist, who wrote:

> *"Evolution is fact. Beyond reasonable doubt, beyond serious doubt, beyond sane, informed, intelligent doubt, beyond doubt evolution is a fact. The evidence for evolution is at least as strong as the evidence for the Holocaust, even allowing for eye witnesses to the Holocaust."*[20]

Evolution in Our Gardens

I live in Canada so a brief winter trip to Florida or some other warm place has been a regular routine for many years.

My first trip to Florida occurred when I was nineteen. This was not only my first trip to this southern state; it was also my first experience with tropical vegetation. I was immediately fascinated with the diverse range of plants that did not exist in my home country. By the end of my first tropical vacation, there were two plants that became symbols for me of this tropical retreat.

First, were the towering palm trees with their trunks and leaves swaying together in the wind. The image of palm trees, often enhanced by their presence along beaches bordering the ocean, was a lingering memory of the tropics. The second symbol was the hibiscus with its large flowers with bright red petals and a brilliant red and yellow pistil and stamen.

Although some time would pass before I successfully kept a palm tree in my home in Canada, the beautiful hibiscus was a different story. For many years I have enjoyed the gorgeous red flowers blooming in my house during the winter and in the backyard in the summer. In the summer, the presence of several hibiscus plants is often enhanced when brilliant colored hummingbirds hover at the mouth of the flowers to enjoy the nectar inside them.

Over time with the help of horticultural experts, an endless array of hibiscus colors has emerged. In addition to the reds, there are now yellows, oranges, white and even a rainbow color comprised of white, orange, yellow and red. As well as the petals evolving into a wide range of colors, the pistils and stamens have also evolved similarly. One can buy a white hibiscus with a red center, or an orange center, or a yellow center, and on the list of varieties go. Although the increasing variety of hibiscus colors is an example of evolution (the process by which living organisms change), it is not an example of natural selection. This is rather an example of artificial selection because the evolutionary changes in these hibiscuses were the result of selective breeding directed by humans.

An example of natural selection can be found in the hummingbirds that seek nectar from the hibiscus flowers. The ancestry of hummingbirds has been traced back some 22 million years.[21] Over millions of years, the beak on the hummingbird became elongated as the result of gene mutations. The reason for this is that hummingbirds with longer beaks were better able to obtain food from inside flowers. Hummingbirds with longer beaks were therefore more adapted for survival. Hummingbirds with longer beaks had an advantage over hummingbirds who had shorter beaks. Over time, the survival of hummingbirds favored those who had the genes that resulted in longer beaks. This is an example of natural selection.

Now at this point, I can hear someone saying, "I understand how natural selection played a role in the development of a hummingbird's beak, but the hummingbird is still a hummingbird—it hasn't evolved into some other kind of animal." Let's address this question.

Sharks Walking On Land

I have mentioned my interest in tropical fish already in this book. Over the years I have kept a wide range of fish, particular marine fish. Two years ago I obtained the egg of a Bamboo shark. While some sharks give birth to live young (viviparous), others lay eggs (oviparous). Approximately 40% of sharks lay eggs. One of the amazing facts about the eggs from a bamboo shark is that the egg case is transparent. This allows one to watch the embryo develop inside the egg case (the egg cases are sometimes called "mermaid purses" because of their shape). It's like having a window into a womb.

Using a plastic clip mounted with a suction cup, I attached the bamboo shark egg on the inside glass of an aquarium. As a family each day, we were able to watch the development of the embryo, from a tiny wiggling worm-like creature until a fully developed six-inch shark pup a little over three months later (the gestation period for a bamboo shark is 14-15 weeks). Once the shark hatched, it became just as fascinating in the aquarium as it was in the egg case.

Bamboo sharks have four pectoral fins (two sets of two) near the bottom of their bodies that are similar to having legs or feet. Although bamboo sharks are proficient swimmers, they also spend a significant amount of time resting on the sand. Interestingly when our bamboo shark was sometimes on the sand at the bottom of the aquarium, it often used its pectoral fins like legs as it strutted along the sand. It also often sat on the bottom with its head raised like a lizard.

It wouldn't take much of an imagination to picture a bamboo shark emerging from the ocean and walking on land like a lizard. After coming to this insight by observing the behavior of our bamboo shark, I was amazed to learn that bamboo sharks can actually live out of water for up to 12 hours.[22] In fact, in Australia these sharks are sometimes called "walking sharks" as they walk from tide pool to tide pool in search of prey. With the beginning form of legs (or feet, depending on how you look at it) and an initial ability to survive out of water, over millions of years I think it's fairly easy to conceptualize a fish like this emerging from the ocean and becoming a land animal, although the fossil record shows that there were other animals that formed in the evolution between fish and land animals.

For example, the 375 million-year-old fossil remains of a creature known as Tiktaalik roseae suggest that this fish-like animal represented an important intermediate step in the evolutionary transition from fish to animals that lived on land.[23] And the flip side of this is that some terrestrial mammals moved back into the water and evolved into whales, seals and manatees.[24]

So yes, there is fossil proof that some animals, over millions of years, actually evolved into completely difference species.

Could All Life Really Have Begun from Just a Single-Celled Organism?

A concern that some people have related to evolution is that according to those who believe in evolution, life began as a single cell (perhaps billions of years ago) and eventually evolved into a creature like us with trillions of cells. Some critics of evolution say that there is no way a single cell could have evolved into a creature made up of trillions of cells, yet this is exactly what has happened to every one of us (and every other creature that has ever existed). A human fetus starts as a single cell and approximately nine months later this single cell has developed into trillions of cells.

Natural Selection, Not God

While natural selection is an incredible marvel, it's fueled by genes that are designed to fight for survival. All life on our planet (and likely in the universe) is impacted by the law of natural selection, and while there are many advantages of this process (without natural selection, we might all still be single-celled organisms), there are some disadvantages. Natural selection has created a war of unbelievable proportions on our planet where every living creature is in a fight for survival, not only with other species, but sometimes even within a species.

In the previous chapter we looked at how many people think that it is God who causes pain and suffering. One of the examples we looked at was the flu pandemic of 1918 where more than 500 million people were infected and somewhere between 50 to 100 million people died. Some people interpreted this huge disaster as a punishment from God (a reference for this was provided in the previous chapter). The truth is that this pandemic was in fact the result of the H1N1 influenza virus.[25]

The 1918 flu pandemic, one of history's deadliest natural disasters, was the result of a viral infection. A virus is a tiny infectious agent that replicates itself inside the living cells of its host. Virus particles are 100 times smaller than a single bacteria cell with the bacteria cell being 10 times smaller than a human cell. Viruses cannot grow or multiply on their own. They need to enter an animal cell (including humans) where like an alien in science fiction movies, they hijack the cell (this process is known as adsorption).[26]

It's estimated that there may be millions of types of viruses, each in a constant battle for survival. In addition to various forms of flu, viruses are also responsible for diseases such as HIV/AIDS and there is growing research linking viruses to some forms of cancer.[27]

If we want to explore the answer to the question, "Why do pain and suffering exist in our world?" we will find our answer in science, not religion. Pain and suffering were not sent by God to punish us. Pain and suffering are largely the result of the scientific fact of natural selection, but without natural selection, our planet would still only be the home of single-celled organisms.

It is natural selection that has led to the incredible diversity of life on our planet, but it is also natural selection that contributes the most to pain and suffering in our world. In the book *The Selfish Gene*, we read:

> *"Natural selection has built us, and it is natural selection we must understand if we are to comprehend our own identities."*[28]

The Answer, My Friend, is Written in Our Genes

Perhaps from your high school biology class you might remember that humans have 46 chromosomes (23 that came from your mother and 23 that came from your father). At the current time it is thought that there are approximately 19,000 genes dwelling on our chromosomes.[29] These genes which carry the instructions for creating each of us have been described by some as blueprints, while others have described them as being more like a recipe.

It is our genes that instruct particular proteins to give us certain traits. Genes are composed of long stretches of chemical building blocks known as our DNA. The complete set of nucleic acid sequences, encoded as DNA within our chromosomes in cell nuclei and in a DNA molecule found within individual mitochondria, is know as the human genome. The first complete sequences of individual human genomes were published on February 12, 2001.[30]

Evolution occurs when genes mutate. While some of these mutations may increase the odds for survival in a living creature, other mutations might contribute to pain and suffering. For example, the long legs and neck of a giraffe were the result of a genetic mutation (or mutations) that favored the survival of giraffes with longer necks over giraffes with shorter necks. The longer legs and neck allowed giraffes to reach higher for food which allowed them to survive at a time when there was greater competition for food at a lower elevation.

In his book *Cosmos*, Carl Sagan writes:

> *"The secrets of evolution are death and time—the deaths of enormous numbers of lifeforms that were imperfectly adapted to the environment; and time for a long succession of small mutations that were by accident adaptive, time for the slow accumulation of patterns of favorable mutations."*[31]

While mutations can aid in the survival of a species, they can also lead to hereditary diseases. Some examples of hereditary diseases are cystic fibrosis, Huntington's disease, and sickle cell anemia. Research also shows that genes can predispose a person to contracting diseases such as cancer, heart disease, and diabetes, although environment factors may also play a role in these illnesses.

In addition, some forms of mental illness may be attributed to our genetic makeup. Throughout history, the mentally ill were often thought to be possessed by the devil. Epilepsy was for thousands of years thought to be caused by demon possession. Throughout much of our history, people with a mental illness were imprisoned and, in some cases brutally tortured as church officials attempted to drive the demons out of a person's body. Even today there are areas in our world where cultures still believe that mental illness is caused by spirit possession, although we know now that this is not true.[32]

Our genes also play a role in our sexual orientation. There is growing evidence that certain gene combinations play a direct role in contributing to homosexuality as an example. Similar to people with a mental illness, homosexuals have suffered in many societies throughout history because religious leaders preached that these people were being controlled by Satan. St. John Chrysostom, one of the early church fathers and the Archbishop of Constantinople in the 4th century, argued that homosexual acts were worse than murder.[33] Early church writings from a variety of sources supported this viewpoint.

Such thinking has continued in many churches into the 21st century even though we now know that homosexuality (as well as other forms of sexual orientation) have their roots in our genes and are not caused by the devil.

In 1993 the so-called gay gene was discovered by Dean Hamer, a researcher at the National Cancer Institute. For men, it appears that homosexuality is caused by a gene somewhere near xq28 along a small stretch of the X chromosome. Gayness may be caused by our genes in the same manner that some people have blue eyes and some people have brown.[34]

Random Pain
Before continuing on to the next chapter, it would be useful to consider how randomness can contribute to pain and suffering. Random events are those that appear to have no order or don't follow some predictable pattern. While the mutation of our genes impacts the process of natural selection, random accidents separate from our genes can cause pain and suffering without appearing to have any direct link to natural selection. For example, a family driving home from an event is hit by a drunk driver, killing all the family members. We can't say that natural selection resulted in this family being killed. Although one might argue that the drunk driver might have had a physiological or psychological predisposition towards drinking, natural selection was unlikely at play in considering the randomness of the time and location that impacted the victims. Unfortunately, the victims were in the wrong place at the wrong time, suggesting that random, unpredictable, unplanned events can occur in our world.

I teach a post-graduate course in counseling. Part of our course looks at career planning. As such, we explore various career theories. One of our class assignments related to this topic is for the students to explain their personal career path in relationship to various career theories.

One of the five theories that is used as a source for this assignment is named the Planned Happenstance Theory. This theory explains how chance events can play a significant role in life choices. Of the five theories we explore in our class, the Happenstance Theory is the one most frequently mentioned by students in their assignments to help explain their career path.

For example, John (not his real name), a high school teacher described how he first pursued a career in accounting. Within a few years on the job, he realized that he had made the wrong career decision. One day a friend who was a high school teacher, asked John to participate as a guest speaker on a career day at her high school. John enjoyed speaking to the students so much that he decided to leave his job in accounting and return back to college to become a teacher.

A chance event led to John going in a different career direction. In addition, once John returned back to college, he met his future wife. In John's exploration of what had happened in his life, it became apparent that a chance event led to both a career change and the meeting of a future wife.

Of course, there are some who would say that it was God who had planned for John to do a classroom presentation that led to a change of careers and the meeting of his wife. For people who believe this, then they must also believe that God was in control of the deaths of 85,000 children under the age of 5 who have died of hunger and disease in Yemen recently.

And then of course, along the same lines, there's the pro athlete who points to God in the sky above after he scores a touchdown or hits a homerun. Or, the interview with a football player after winning the Super Bowl where the player thanks Jesus for helping him. One might ask such athletes, where were Jesus and God when 85,000 children in Yemen really needed some help? Are we really so self-absorbed in the Western world that we think that God is involved in helping football players win games while he ignores helpless children who are dying in other parts of the world?

Chance events can result in happy endings, although they can also contribute to pain and suffering.

How many people died in the twin towers on September 11, 2001, who didn't even work there? How many people on that tragic morning were attending a meeting or a job interview, who normally wouldn't have even been in the building? And how many people were supposed to be in one of the two towers on that morning, but got tied up in traffic or were late arriving because of a family concern? And if you believe that it was God who saved those people whose plans unexpectedly changed that morning to prevent them from being in the twin towers, then by the same logic you must also believe that it was God's plan to have everyone inside the towers slaughtered.

Chance: it happens every second, every day to all of us, and it can result in pain and suffering, but it can also result in success. In his book *The Drunkard's Wife*, Leonard Mlodinow, a Caltech professor who teaches about randomness, writes:

> *"We're continually nudged in this direction and then that one by random events. As a result, although statistical regularities can be found in social data, the future of particular individuals is impossible to predict, and for our particular achievements, our jobs, our friends, our finances, we all owe more to chance than many people realize."*[35]

Although any statistician will tell you that you are wasting your money when you buy a lottery ticket, every week another person becomes a millionaire because they bought the ticket against all odds. Random events occur every day in every country and affect every person. Some might describe random events as luck, whether good or bad. Others might use words such as happenstance or serendipity. Whatever word might be chosen to describe these events, they are still random rather than preordained.

Earthquakes and Miracles

An earthquake brings a huge apartment complex to the ground. Rescuers struggle to find survivors. Eventually, the news media will find the story of one person who has "miraculously" survived and is still trapped in the rubble.

For hours and sometimes for days the media will focus on this rescue attempt, often using the word miracle to describe what is happening. In such a tragedy, hundreds (and sometimes thousands) will perish. A real miracle might be if the building collapsed and every single person survived, which of course never happens. On the other hand, blind chance would dictate that one person, out of the hundreds, just happened to be in the right place at the right moment to escape being crushed.

Random events don't just impact individuals; they can also occur on a much larger scale. Consider the death of the dinosaurs. Most scientists agree that a random asteroid struck our planet about

65 million years ago causing an environmental catastrophe that ended the era of the dinosaurs (although it should also be noted that there are some scientists who believe that a series of erupting volcanoes might have caused this disaster).[36] While it could certainly be argued that the animals that survived this random tragedy were favored by natural selection, it could also be said that the dinosaurs were exterminated by a random event (regardless of whether it was caused by an asteroid, or erupting volcanoes).

In summarizing this chapter, we have looked at how natural selection, rather than God, is the greatest cause of pain and suffering on our planet. We have also considered how random events can cause pain and suffering. In the next chapter, we will take a further look at the implications of natural selection for humans, especially in relationship to examining the problem of pain and suffering.

- 4 -
Are Pain and Suffering Different for Humans?

In the previous chapter, we looked at how the law of natural selection results in a fight for survival throughout our planet. No animal on our planet is free from the implications of natural selection, and that includes us. Although pain is largely a byproduct of the battle for survival, there are some differences—both positive and negative—that impact humans more than most other animals. In this chapter, we will explore how natural selection has impacted pain and suffering for us as humans.

Freedom from Pain
Congenital analgesia is a rare condition in which people cannot feel physical pain. People with congenital analgesia are often at serious risk from both disease and injuries. Most people, upon touching a hot stove, would immediately remove their hand. Someone with congenital analgesia might suffer serious burns in a similar situation because they don't feel any pain, therefore they don't pull their hand away. Likewise, someone with congenital analgesia might be sick, but may be unable to recognize this because they can't feel any pain.

I think you can see that people with congenital analgesia are constantly at risk of further injury, illness or even death because they can't feel pain. For those of us who do not have this genetic condition, in spite of any concerns we might have about experiencing pain, we should be able to understand that pain can often protect us from further injury as we instinctively flee (or fight) whatever is causing the pain. As Professor Donald M. Broom of the University of Cambridge states:

"The advantages of pain are that action can be taken

> *when damage occurs, consequent learning allows the minimizing of future damage and, where the pain is chronic, behavior and physiology can be changed to ameliorate adverse effects."[1]*

Do All Lifeforms Experience Pain?

Pain is part of the body's defense system which protects us from further harm and aids in our survival. It is generally accepted that any animals that do not have nociceptors (receptors for pain) cannot feel pain in the same sense that we do, but just because an animal—let's say a single-celled organism as an example—doesn't feel pain, this doesn't mean that they don't react instinctively to unwanted stimuli.

For example, single-celled organisms that are able to move will retreat from noxious stimuli while those that are unable to move may enter a dormant state (called endospore) where they may wait for better conditions before they continue their growth.[2] Similarly there are some studies that demonstrate that even plants can respond to negative stimuli. In the book *The Intelligent Plant*, we read:

> *"Unable to run away, plants deploy a complex molecular vocabulary to signal distress, deter or poison enemies, and recruit animals to perform various service for them."[3]*

The ability to feel pain, or instinctively react to it, is a survival mechanism that has developed throughout our evolutionary history. Although many animals experience pain, it is generally believed that there is a significant degree of difference in the amount of pain that various animals feel. For example, a dog appears to feel pain much more severely than let's say a worm. The presence of a suitable nervous system and sensory receptors—that exist in a dog, but not a worm—increase the amount of pain felt.

Animals may demonstrate their pain through a lack of eating, a change in behavior, respiratory changes, inflammation and cries of distress. It should also be noted, though, that just because an animal doesn't cry out doesn't necessarily mean it isn't experiencing pain. In the wild, it can be a survival advantage for an injured animal to be stoic and keep quiet, and this survival mechanism may still be in the genes of some of our household pets (and perhaps even in some of us, particularly men who throughout our history have often been the hunters and warriors; there may have been a survival benefit for men not to cry out in pain when they were wounded during a hunt or battle). It can also be a survival mechanism in the wild for some animals to resist showing any symptoms of pain, otherwise they might lose their place in the social hierarchy of their group.

Pain and Brains

It would appear that the more developed the brain is in the evolutionary scheme of things, the more troubling pain can be. For example, insects who lack nociceptors (which are directly responsible for feeling pain) seem to be able to go about their daily tasks without expressing pain even if they are missing a limb.

Our larger brain and more developed nervous system have resulted in us experiencing pain to a greater degree than most other animals (although we might also experience pleasure to a greater degree than most other animals). In addition to suffering physical pain, we have the added problem of being affected by emotional pain. As J.K. Rowling writes in *Harry Potter and the Order of the Phoenix*:

> *"I DON'T CARE!" Harry yelled at them, snatching up a lunascope and throwing it into the fireplace. "I'VE HAD ENOUGH, I'VE SEEN ENOUGH, I WANT OUT, I WANT IT TO END, I DON'T CARE ANYMORE!"*
> *"You do care," said Dumbledore. He had not flinched or made a single move to stop Harry demolishing his office. His expression was calm, almost detached. "You care so much you feel as though you will bleed to death with the pain of it."*

Recent discoveries have revealed that the brain pathways of primates (monkeys, apes, humans) evolved differently than those of other animals.[4] Although pain pathways are present in all mammals, an additional pathway emerged in primates at some point in our distant past that directly linked the spinal cord with the nucleus of the thalamus activating the anterior insula and the anterior cingulate cortex, resulting in the connection of pain to our emotions.[5] In other words, while pain might be purely a physical response in most animals, in some animals pain also has a significant emotional component.

Emotional Pain

For humans, the emotional component of pain (and on the flip side, the attempt to maximize pleasure) has become an integral characteristic of our species. While chronic pain can have a physical cause, studies have also shown that it can also be the result of stress and emotional issues.[6]

Researchers at Perdue University in the U.S. and the University of New South Wales in Australia found that pain caused by emotional distress is more deeply felt and longer lasting than that by physical injuries. With more developed emotional centers in our brains, we have the potential to suffer pain at a more profound level than most other animals.[7]

The same areas of the brain (the anterior insula and the anterior cingulate cortex) get activated for both physical and emotional pain.[8] Although we sometimes talk about physical and emotional pain as two separate things they are so connected that physical pain can be exacerbated by our emotional response to it, and similarly what began as emotional pain can result in an actual physical issue.

In her book *The Difference Between Pain and Suffering*, Catherine Carrigan writes:

> *"We can be so heavily invested in our tales of sorrow and woe that we have no clue how much repeating that negative story creates the chemistry in our body mind that triggers the physical pain."*[9]

In other words, some people who continue to dwell on feeling sick or experiencing pain, even if there is no physical cause there in the first place, can eventually suffer pain daily as though they actually had an illness or physical ailment.

In addition, when your emotions become involved in an actual illness that you have, this can prolong the illness. For example, you get the flu and as a result you miss an important event that you were looking forward to. You become so emotionally distraught that you missed this event that this then interferes with your efforts to overcome your flu. Your flu (that might have subsided in a few days) lasts for a week or more. Similarly, a person who you care about deeply unexpectedly ends her relationship with you. You are so emotionally crushed that your body's immune system is lowered and you get physically sick. Emotional pain can hurt you physically and physical pain can hurt you emotionally.

Feeling the Pain of Others
Another distinction of pain for humans is our ability to feel the pain another person is suffering,[10] although it is quite possible that some other animals also feel empathy towards others (both humans and non-humans) who are encountering pain (and it might be added that we may also feel the pain that one of our pets is experiencing). Certainly, many dog owners believe that their pets can sense suffering in other family members. Award-winning environmental writer Carl Safina states:

> *"Many people think that empathy is a special emotion only humans show. But many animals express empathy for each other. There are documented stories of elephants finding people who were lost. In one case, an old woman who couldn't see well got lost and was found the next day with elephants guarding her. They had encased her in sort of a cage of branches to protect her from hyenas."[11]*

Empathy as expressed through words, both written and verbal, as well as through body language (especially touch) can also cause pain for the person who is being empathetic. Research shows that the closer the relationship between two people, the greater the pain the empathizer may experience.[12]

The Pain of Loss
Loss, such as breaking up with a significant other, and pain

caused by events such as rejection or bullying can all contribute to suffering.

I once counseled a woman who had experienced painful injuries from a car accident, a serious illness, and most recently the loss of a job that she loved. In her words, her recent job loss was causing her to suffer more than any of her other experiences with pain even though this did not have the direct physical component that her other painful experiences held. Emotional pain can be every bit as powerful as physical pain.

Some people never recover from the death of a loved one. Grieving in humans can be a significant source of pain, and one that without help can become debilitating.

Do Animals Grieve?

Pain is a complex experience involving both sensory and emotional components. It's possible that similar pain exists in some animals, other than humans. In the book *How Animals Grieve*, author Barbara King describes how after the death of a female elephant, her body was visited by females from five different herds. During these visitations, the elephants rocked back and forth while standing over the body.[13] In this same book King talks about other species of animals who appear to mourn their dead. Marc Bekoff, professor of Ecology and Evolutionary Biology at the University of Colorado, states:

> *"Many animals display profound grief at the loss or absence of a close friend or loved one."*[14]

In the summer of 2018, an orca whale off the west coast of Canada carried the body of her dead calf on her back for at least 17 days. Deborah Giles, a biologist who studies killer whales stated, "This is an animal that is grieving for its dead baby, and she doesn't want to let go. She's not ready."[15]

So yes, it's quite possible that some non-humans can express both empathy and grief, but for now we're going to focus on the human species.

Language and Pain

Although some animals in addition to humans appear to express emotional pain such as grief and/or sadness, for humans these are known and established facts. As well, the pain we experience may be further increased both in immediate severity and duration by our thoughts and use of language.

Many animals can communicate with each other through sounds. Think of a bird singing, an act that might signal danger or even availability for mating. Or a dog barking, once again possibly signaling danger, or even demonstrating affection for a companion. Animals express a wide range of sounds in their ability to communicate with each other, but of all the animals we stand alone in the sophistication of the use of language.

For example, we can even experience pain or happiness through reading words, something that no other animal is able to do (although some dolphins have been taught to read two-dimensional symbols, and Washoe, a chimpanzee, learned to communicate using American sign language, as did Koko the gorilla, but in these examples it could be argued that the animals were responding to visual symbols rather than reading actual words).[16]

What other animals, besides humans, can read a story and cry, or watch a movie and have a strong emotional reaction to various scenes?

Bad News Sells. If It Bleeds, It Leads.

We can experience pain through the written word or through artificial stimuli (such as television shows or movies). News articles of tragedies in our community or around the world can be painful to read or watch. Animals other than humans are unable to read about tragedies that have occurred elsewhere. While a pet dog might be aware of something happening to another dog in the neighborhood, this dog isn't constantly exposed to what is happening to other dogs in communities throughout the world. For humans, the flow of negative news is constant. While a wild animal might observe another animal being killed (or even be involved in the killing), this same wild animal does not witness killing after killing in the manner we are exposed to both in our news and TV shows.

For humans, this constant bombardment of negative news can

cause emotional pain that is every bit as real as physical pain. Some psychologists believe that exposure to negative and violent media can cause serious psychological concerns.

The work of British psychologist Dr. Graham Davey, suggests that violent media can exacerbate or contribute to the development of stress, anxiety, depression and even post-traumatic stress disorder.[17] This is a concern that impacts humans more than other animals (when was the last time you saw a dog surfing the internet or sharing tweets?).

A study involving 32,000 participants by psychologists at the University of Liverpool found that how people think about traumatic events is a key factor in contributing to anxiety.[18] It's not necessarily the event itself that triggers the anxiety or depression, but rather what we actually think about the event.

For the past few decades I have taught university courses in counseling, primarily to teachers and social workers. As we explore the role of counselors and social workers in schools, my students have over the years increasingly talked about the increase in anxiety among teenagers.

The perception of the teachers and social workers is that students today are being bombarded with negative news that is being brought to them instantly and graphically. They report that teenagers even say things like, "When I go into a crowded building, the first thing I look for is an exit in case a shooter appears," or "What's the use of making any future plans when the world's falling apart?"

In the UK, a government report found that teenagers are suffering record levels of anxiety.[19] The same study showed that girls were particularly negatively affected with 37% of them having symptoms serious enough to see a doctor.

As humans, our pain and suffering are increased by readily listening to, watching, or reading about negative news or troubling fictional stories that are conveniently brought to us through various devices instantly and 24 hours a day, 7 days a week.

News and some fictional stories can cause us to worry irrationally about the coming of a nuclear war, the growing threat of superdiseases, or the impact of climate change, just to name a few concerns. While animals such as fish or coral may suffer the devastating impact of climate change, they don't have the added pain of being

PAIN & SUFFERING FOR HUMANS / 83

emotionally flooded with troubling climate change reports or predictions. As reported in the *New York Times*, "Bad news sells. If it bleeds, it leads. No news is good news, and good news is no news."[20]

Which Would You be More Likely to Read or Watch: Positive or Negative News?

Some studies confirm that we are more attracted to negative news than positive. In addition, it has been shown that our brains are more sensitive to negative triggers than positive ones.[21] There is a tendency for us to be more fearful than happy when we watch the news, which in turn can cause pain or magnify any pain that we might already be experiencing.

Are My Friends Happier Than Me?

In addition to the pain that we might experience from observing or reading about tragedies throughout the world, our well-developed language enables us to share the details of what is happening in our lives with others in so many easy-to-use formats.

We can Skype, we can email, we can text, we can tweet, we can share both our pain and happiness on Facebook or Instagram, and so on. This ease of sharing spreads our experiences at an incredibly rapid rate among our friends (and even among strangers).

Even when others share positive news on social media such as Facebook (think of people who post pictures of their gorgeous vacation), the longer that people spend on Facebook, the more they think that life is unfair which can then result in unhappiness and emotional pain.[22]

German researchers who found similar results among Facebook users described this as "the self-promotion-envy spiral."[23] In this study, the German researchers found that about a third of the 600 adult participants involved in the study said they experienced mostly negative feelings when using social media. These studies show that even when the news appears to be good (at least to the person posting it), some people will still be affected adversely.

Past Painful Experiences

Another aspect of pain that we experience as humans relates to our thoughts about past painful events or even the anticipation of future painful events.

For example, if you dislike going to the dentist and you have an upcoming procedure to be done, you might find yourself experiencing pain days before the actual appointment. This pain might even grow in intensity until the day of your appointment, and it's even possible that you might continue to dwell in a negative manner on this appointment days or even weeks after it's over.

There is some debate whether any animals other than humans can experience pain by thinking about either the past or future. As you might imagine it's very difficult to actually measure what animals, other than humans, are thinking. While there is some research that demonstrates that some non-humans have what we would call consciousness or an awareness of self, it's difficult to ascertain whether this consciousness includes past or future thinking.

While some animals may learn behavior that is based on past experiences this doesn't necessarily mean that they can consciously think about their past. Their behavior might simply be a learned response to a past stimulus that is now repeating itself. When you pick up your dog's leash and it runs to the door, wagging its tail, this might be the result of a conditioned response rather than the result of consciously thinking about a similar past experience. And it should be noted that a negative reaction to a dentist appointment can also be a learned response for a human, but this response can be exacerbated by our thinking about it.

Thinking about traumatic events from the past can cause various degrees of discomfort in the present. For some, a dentist appointment might result in sweating, a lack of sleep, headaches and even toothaches. For most people these concerns subside after the appointment. There are though other forms of traumatic events that can cause crippling pain for weeks, months, and even years after the event.

PTSD

A serious problem regarding soldiers returning from war zones is post-traumatic stress disorder (PTSD). Some soldiers find themselves visualizing the horrors they witnessed during their combat missions even though there may not be any related stimuli triggering such thoughts (although loud noises or visual cues on TV or in movies may trigger this response). These thoughts can result in debilitating emotional and physical pain.

The unfortunate reality is that PTSD can impact anyone who has suffered a traumatic event in their past. This can include sexual assault victims, other forms of abuse, car accidents, shootings, natural disasters and so on. I was recently told by a teacher that there were students in her school who she believed suffered from PTSD because of the amount of violence that occurred in their school community.

Unfortunately, PTSD can lead to further problems causing pain. Research findings show that people with PTSD are at a higher risk for suicide and intentional self-harm.[24] In the U.S. about 3.5% of adults have PTSD in any given year, and about 9% of people in the U.S. will develop it at some point in their life.[25] Even when we don't label a reaction to trauma as PTSD, these events, whether real or imagined, can cause troubling pain. And for some, the impact of watching troubling news or graphic violence in movies or in video games might trigger responses that are similar to PTSD.

After the kidnapping, sexual assault, and savage murder of a teenager in my community, as a crisis counselor I worked with some teenagers who were friends of the victim and others who were impacted by the tragedy even though they didn't personally know the victim. These teenagers (none of whom actually witnessed the attack), suffered overwhelming anxiety, sometimes even months after the horrific tragedy, because they either reflected on whether the attack could have happened to them, or they thought about the possibility of a similar attack happening to them in the future.

After the murdered girl's funeral, I was asked to present a workshop for the staff at her high school on how to help their students deal with what had happened (specifically focusing on looking at the why questions that their students were asking). This was one of the most difficult things I have ever had to do.

The following year when another teenager was kidnapped, sexually abused, and murdered, fear spread among female teenagers in our community and beyond. At this time, the impact on teenagers in our community was so strong, I was asked to present several workshops to other counselors and social workers on how to help teenagers during this difficult time. As a result of these workshops, I became even more aware of how these horrendous events were impacting teenagers (and parents).

Many of our teenagers, especially girls, couldn't sleep at night. Some of them had nightmares of being attacked. Some slept with knives under their pillows. Others carried knives in their purses when they went to school. Many of these students had trouble concentrating on their schoolwork, and some of them experienced headaches and vomiting.

Unfortunately, real-life stories such as these often become common themes both in our news and in TV shows, books, video games, and movies. As a result, young people (and adults) can suffer PTSD symptoms even when they don't personally know the victims, and even when the stories might be fictional.

Pain itself, or even the mental fabrication of pain, can cause further pain. Chronic pain, which is often defined as pain that has lasted for 3 to 6 months or longer, can cause sleeplessness, anxiety and even depression. Our brain can even rewire itself to perceive pain signals even if the signals aren't being sent anymore.[26] In other words, we may continue to suffer pain even after the physical or emotional stimulus that first caused the pain is no longer present.

Summary, So Far

Let's take a moment to review what we have looked at so far in this chapter. It appears that humans experience pain more severely than most other animals because there is an emotional component to our perception of pain (and because we have more developed pain receptors). In addition, we may experience some forms of pain through various forms of the media that other animals do not experience. As well, the manner in which we think about any pain that we might be experiencing can impact the amount of suffering that occurs.

Is There a Positive Side to Pain?

As explained earlier in this chapter, pain can protect us. We remove our hand from a hot stove because pain tells us to do so. Both physical and emotional pain can protect us by motivating us to remove ourselves from a painful experience (whether physically or mentally).

It's also useful to consider that pain in humans does not always cause suffering. Athletes may push themselves well beyond any pain threshold in their training but may not experience suffering as a result of the pain. In fact, an athlete in such a situation might even experience a sensation of pleasure. Perhaps you've heard of the runner's high. For some runners the pain and exhaustion that might be a normal reaction to a lengthy run are replaced by a feeling of euphoria.

Researchers in Germany have found that running elicits a flood of endorphins in the brain causing a sense of euphoria.[27] Dr. Henning Boecker, the lead researcher in this German study, found that there were stories of runners who had experienced stress fractures, even heart attacks, and kept running. There are certainly verified stories in professional sports of athletes who kept competing or playing even after suffering what might be considered a game-ending injury.

Overcoming Pain

Baseball pitcher Curt Schillings played through seven strong innings in the sixth game of the American League Championship Series in spite of playing with a torn tendon in his ankle. Bryon Leftwich, a quarterback for Marshall college, broke his left tibia during a game in 2002 and then returned to lead his team to a 17-point comeback. Toronto Maple Leaf hockey player Bobby Baun broke his ankle in the sixth game of the Stanley Cup finals and was carried off the ice on a stretcher. Baun then returned in overtime to score the winning goal. Shun Fujimoto broke his kneecap during the men's team gymnastics competition in the 1976 Summer Olympics. In spite of his injury, he competed in the last three events, helping his team win a gold medal. And then there's Kerri Strug competing on the American women's gymnastic team in the 1996 Summer Olympics. Coming to the final event Strug needed a vault score of 9.5 or better, but on her first attempt she tore two ligaments in her ankle. In spite of this injury, Strug vaulted again, landing on one leg, and scoring 9.712 to help win gold for her team.

Stories of people overcoming pain are not limited to sports. In everyday life there are people who manage pain to achieve success.

I'm sure that almost everyone reading this book can identify someone who fought through a serious illness such as cancer to live a meaningful life, even if the disease in some cases eventually killed them. Helen Keller, who at nineteen months of age contracted an unknown illness that left her blind and deaf, only to eventually become an activist and acclaimed author, said:

> *"All the world is full of suffering.*
> *It is also full of overcoming."*

While pain may have a greater impact on humans than other animals, our well-developed brain also gives us the ability to reduce pain through our thinking. Due to our language ability, we are able to better describe what is hurting us and obtain more appropriate medically prescribed relief for our pain. When it comes to pain, our highly evolved brains can be both a curse and an advantage.

There are multiple ways that people can use their minds to cope with or reduce pain. These can include relaxation, meditation, and positive thinking although there are other mind-body techniques as well. Various therapy techniques such as cognitive-behavioral therapy may also be useful.[28] Cognitive behavioral therapy helps people to find and practice effective strategies to decrease the symptoms of disorders.

There are also new techniques such as real-time functional magnetic resonance imaging (rtfMRI) which appear to hold promise by helping people to take control of specific regions of the brain to better control pain. The initial research with rtfMRI has shown that those who exercised the greatest control over their brain activity were able to experience the greatest benefit in pain reduction which brings us back the point that our thinking can have a strong impact on how we perceive our pain.[29]

In considering how our thinking may affect pain, we might also ask how prayer might impact the reduction of pain and suffering.

It is estimated that in the U.S. more than half the population pray to some degree every day. Research shows that prayer can have some benefits to the person praying. Prayer on a regular basis can improve self-control which can in turn help a person to make better choices. Research has shown that prayer, especially when the prayer focusses on the welfare of others instead of praying for material gain, helps to eliminate stress which can assist with controlling or reducing pain.[30] Self-prayer has some similarities to meditation or positive thinking, and as such it might be a useful strategy for some in coping with pain.

On the other hand, research into the act of praying for others has shown that prayer did not heal illness in others. In fact, when people who were about to face surgery knew that others were praying for them, they actually had a higher rate of post-operative complications. This study involved more than 1,800 patients and has been conducted for more than a decade.[31]

The development of our sophisticated use of language (both in written and verbal forms) can contribute to pain (as discussed earlier in this chapter), but it can also contribute to overcoming pain. In a review of David Biro's book, *The Language of Pain: Finding Words, Compassion, and Relief*, the reviewer writes:

> *"During moments of crisis—including illness and suffering—stories are the way we human beings make sense out of chaos, meaning out of madness, and form relationships critical to our survival and healing."*[32]

In considering the various "thought" options that might be available to assist with pain reduction, the intent of this book is not to be a manual on using the mind to deal with pain. I would simply like to establish the fact that our thoughts can help us to cope or even reduce any pain that we experience, although as explained earlier in this chapter our thoughts can also exacerbate our pain.

Although natural selection has given us a brain and nervous system that results in us feeling a heightened sense of pain, these same physical characteristics allow us to experience a heightened sense of happiness and pleasure (more on this in Chapter 11). In a U.S. National Library of Medicine article titled *The Neuroscience of Happiness and Pleasure*, we read:

> *"The evolutionary imperatives of survival and procreation, and their associated rewards, are driving life as most animals know it. Perhaps uniquely, humans are*

90 / WHY

*able to consciously predict and anticipate
the elusive prospect of happiness."[33]*

Aristotle argued that happiness depended on hedonia (pleasure) and eudaimonia (a life well lived). A life well lived might also be interpreted as living a life with meaning (which is the theme of this book). A detailed exploration of our experiences with pleasure would find that the same explanations for experiencing pain that we have looked at in this chapter (the vast neural networks in our brain and nervous system) are also responsible for our feelings of pleasure.

In the next chapter we will explore the question of whether there is life after death, which has been seen throughout history as the ultimate escape for the suffering that one might experience here on earth, although some also see eternal life as the ultimate reward for pleasing the God or gods of their religion.

- 5 -
Is There Life After Death?

Years ago, a husband and wife, in their early thirties, came to me for counseling. Their concern, as they began to describe it, focused on their eight-year-old daughter. Let's call her Jane.

Jane was born with a number of medical concerns in addition to very low functioning mental ability. Although Jane was able to walk, she had never learned to speak other than a few guttural sounds. Jane's parents were very religious, often attending their fundamentalist church several times on Sunday in addition to a few times during the week. Even though they generally took Jane to church with them, they expressed their thoughts to me that they were certain that Jane didn't understand anything about God.

As our conversation continued, Jane's mother came to the agonizing crux of their current concern. Jane had recently been diagnosed with terminal brain cancer. At best, she might have another few months to live. Both parents had accepted that Jane was going to die. What was tearing them apart was their concern about whether Jane would go to heaven.

Through their Christian religion, Jane's parents were familiar with Bible verses that promised eternal life for people who believed in Christ. Unfortunately, they said painfully that Jane didn't have the mental ability to believe in Christ. These distraught parents were crushed to think that their daughter might not go to heaven because she was incapable of understanding the concept of salvation through Christ. Jane had suffered enough on earth; surely, they hoped, she would be rewarded with peace in heaven.

Over the years as a counselor and as a member of a relatively large family, I have attended a significant number of funerals. Whether the deceased were two years old or ninety-two years old, there is no doubt in my mind that most family members and friends found comfort in the belief of life after death, so I am reluctant to say anything that might diminish this belief, but I am also aware that this same belief, that is a key concept in the major religions of the world, has led to the deaths of millions through wars and persecution, and caused horrific suffering.

Suicide Bombers

The Islamic extremist suicide bomber who blows himself up to kill dozens (or hundreds) of innocent victims has a strong belief in life beyond death. In August of 2002, CBS aired an interview with Muhammad Abu Wardeh who helped to recruit terrorists for suicide bombings in Israel. Abu Wardeh was quoted as saying:

> *"I described to him (the recruited terrorist bomber) how God would compensate the martyr. If you became a martyr, God will give you 70 virgins, 70 wives and everlasting happiness."*[1]

Significant death and suffering have occurred and continue to occur as a result of the belief in life after death. Without this promise of an inviting paradise, we might very well ask if these Islamic men would so willingly blow themselves and others apart, or if Muslims or Christians would have so readily sacrificed their lives in many "holy" wars throughout history.

Extreme Muslims are not alone in their willingness to sacrifice their lives for their faith. One of the most frequently quoted Bible verses provided in honoring soldiers who have died on the battlefield is John 15:13 - "Greater love has no one than this: to lay down one's life for one's friends." Add to this, biblical verses such as Revelation 21:4 - "He will wipe every tear from their eyes. There will be no more death or mourning or crying or pain, for the old order of things has passed away."

The promise of an everlasting paradise is a powerful concept in religion. While on one hand it can be the source of significant comfort for those who are fighting a serious illness or for those who have lost a loved one, the other side of this is that the promise of a paradise can inspire soldiers to sacrifice their lives.

IS THERE LIFE AFTER DEATH? / 93

Unhappiness today, heaven tomorrow
Another concern related to the concept of heaven is that a significant number of people with strong religious beliefs are so focused on heaven that they sometimes fail to enjoy every day here on earth. Such thinking can also lead to apathy in dealing with current issues, whether these concerns are personal or whether they are occurring on a more global scale such as climate change (we will look at this in more detail in Chapter 9).

Similar to previous chapters, let's look at the historical context of how we reached our current views regarding life after death. After doing this, we can then consider how these views might be affected by new learning in the 21st century.

Reverend Jim Bakker, a master of commercializing evangelical Christianity, stated:

> *"We have a better product than soap or automobiles. We have eternal life."*[2]

Why Do We Bury Our Dead?

Let's begin by travelling back a few hundred thousand years. The presence of gravesites, or deliberate burying of the dead, is generally thought to provide evidence that the people who did this had some form of belief in life beyond death, even if we don't know the full details of such a belief.

One of the earliest possible deliberate graves might have occurred in a cave at Atapuerca, Spain, where over 6,500 hominin fossil bones and teeth (representing about 30 individuals) were found in a chamber that was named Sima de los Huesos (Pit of the Bones). These remains were initially dated to 250,000 years ago but have recently been moved back even further to around 430,000 years ago.[3] In this particular site, there was also the finding of a hand-axe made from red quartz. The presence of the axe might represent a very early belief in providing a weapon or tool for the departed to use in the next life, but obviously there is no way to completely substantiate such a claim.

Moving on to around 100,000 years ago both Homo sapiens (that's us!) and Homo neanderthalensis (our cousins) began to bury their dead in caves and rock shelters on a consistent basis which might suggest they were trying to protect the dead for their journey to an afterlife.

Evidence of this was found in the Qafzeh and Es Skhul Caves in Israel, where human skeletons were discovered between 1929 and 1935 (the Skhul remains having been dated to 100,000-135,000 years ago using the thermoluminescence technique as well as the electron spin resonance technique). The discoveries also included bones that were stained with red ochre, providing proof that some form of symbolism was associated with the burials (some Indigenous communities in Australia still paint the bones of their dead with red ochre as part of their funeral rituals, although this doesn't necessarily mean that this ritual is the same as what occurred 100,000 years ago). In addition, tools and sea shells (some also having red ochre on them) were also found in the caves which might also represent some form of burial ritual.[4]

It is generally thought that the burial of the dead represented an early form of religious practice, a practice that demonstrated a belief in what happens after one dies. The presence of tools and weapons in some of these early burials suggest that our ancient ancestors developed a belief system that death was not the end of the road, a belief that the religions of today have maintained.

The origin of such beliefs often had their roots in our ancestors observing their surrounding world. For example, as early humans observed plants dying in the winter, only to miraculously resurrect in the spring, it's not a stretch by any means to consider that the burial of the dead was similar to a plant withering in the ground in the winter, with the hope of being resurrected in the spring. Another possible rationale for the burial or the hiding of the dead might be that our ancestors believed it was important to protect the bodies of the dead (perhaps because they believed in the body having a future life beyond death).

For most of human history, lives have been short with brutal suffering occurring on the level that many animals experience in the wild. It was rare for people to live beyond their twenties. With such short lives where pain and suffering were daily common experiences, it is no wonder that humans who had the intellect to think abstractly would hope and believe in the possibility of life after death.

Do We Have Souls?

A fundamental aspect of believing that life exists after death rests in the belief that we have a soul. This concept can be traced back to the Egyptians and most likely well before the Egyptians.

Egyptians thought that individuals were made up of both physical and spiritual elements. While it is likely that some humans before the Egyptians shared similar beliefs, the Egyptians have provided us with actual historical records to explain their beliefs.

From Egyptian hieroglyphics we have learned further details about their belief in immortality. The Egyptians believed in a vital essence in each person known as ka. Ka entered a person's physical body at birth, and after death the physical body would become a vessel for the ka (which is the reason why it was so important to use mummification to preserve the physical body after death). If the body decomposed, the ka could die. The Egyptians also believed that a person's personality (known as ba) could be reunited with the ka after death (this unification was known as akh). If a person's ka failed to reunite with their ba, then they would suffer eternal death (but not eternal punishment).

Similar to the journey of the Egyptian god Osiris, it was believed the ba would return to the mummified body at night and rise like the sun again in the morning. While some experts view this reunification as having an eternal physical component (thus the reason for the burial of food, tools and even servants along with the dead), others see this as more of a spiritual presence that could interact during the day with the gods outside of the tomb.[5]

As we move on to the Greeks and Romans, we continue to see evidence of a belief in the duality of a person which includes both a soul and a body, a belief that once again has been borrowed by the major religions of today's world from the distant past.

Do All Animals Have a Soul?

Some religions (or sects within certain religions) teach that all living creatures have a soul, although this soul is generally not considered to be the same as the soul in humans. It is usually accepted (within many religions) that the soul in animals, other than humans, is the animal's life force while the soul in a human is immortal.

For those who believe that we have a soul, it could be asked at what point in the evolution of humans did God decide to give us a soul that was immortal? And to this, we might add, what was the criteria used by God to decide that humans, out of all the creatures on this planet, were to receive immortal souls?

To answer such questions, we need to explore how humans might be different than other animals.

A few decades ago, humans were defined as creatures who used tools. The problem with that definition (and the reason this definition has now been discarded) is that there are other animals who also use tools.

For example, for thousands (perhaps millions) of years chimpanzees have used tools, often making stone hammers or creating spears to use in hunting, or even using specialized tools to hunt for ants. Crows are able to craft twigs and even their own feathers into tools. Elephants can modify branches from trees to swat flies away. Dolphins can carry marine sponges in their beaks to uncover prey in the bottom sand while protecting their beaks. Octopuses can use discarded coconut shells for protection.

Finding a definition of human is not easy. Most dictionary definitions state that humans are human beings or are people. This doesn't really help us because the question still remains, "At what point in prehistory did humans separate from other animals by gaining an immortal soul?" If we are unable to answer this question, we might have to accept that all other animals must also have an immortal soul, or maybe we might have to conclude that there is no such thing as an immortal soul.

And if all animals have a soul, it must follow that the immortality of the soul has nothing to do with religion because no other animals, other than humans, have any awareness of religion.

What Does Our DNA Tell Us?

Somewhere along the way in a school textbook, you might have seen a diagram showing various forms of apes evolving into humans. In this diagram, there is a usually a straight-line progression of hunched-over ape-like creatures progressing in each picture until they are standing a little more erect (and a little less hairy, and a little more human in each picture). Such a diagram is no longer scientifically accurate.

IS THERE LIFE AFTER DEATH? / 97

We now know that there have been up to a dozen varieties of modern humans who have lived during the past 2 million years.[6] The most familiar of our cousins would be the Neanderthals, often represented in movies and cartoons as cave men. The Neanderthals provide evidence that the concept of humans having a soul is not as straightforward as some people might like to believe.

The Neanderthals have often been viewed as being more ape-like than human. Recent scientific evidence shows that this is false. Using DNA studies, we now have conclusive proof that our most direct ancestors interbred with the Neanderthals. It has been found that modern Europeans and people of European descent have 1.8-2.6% Neanderthal DNA in their genomes.[7] This supports the fact that not only did our direct ancestors have sex with the Neanderthals but their offspring were fertile which means that Neanderthals and us were part of the same species (animals of differing species can have sex, but they can't have fertile offspring).

DNA studies have also shown that Neanderthals and the Denisovans (another early form of Homos) also interbred. And DNA evidence has also shown that Denisovans and Homo sapiens interbred. In the southeastern part of our world modern humans have 3-5% of their DNA derived from the Denisovans.[8]

What does this mean as we consider the concept of a soul? Well it means that if modern humans have a soul then our ancestors from hundreds of thousands and even millions of years ago must have also had a soul. In looking at the evolutionary history of humans, there is no sudden departure from other early forms of humans which might indicate the perfect opportunity for God to insert an immortal soul into Homo sapiens. In fact, the DNA evidence shows that we interbred with early forms of the Homo genus, even though they may have been more ape-like in appearance than human.

Let's take this back a step further. The fossil record indicates some links between the Homo genus and human-like apes. If all early forms of humans had a soul, then it would follow that the ancestors of these humans and their relatives must also have had a soul. In looking at our evolutionary past, there is no clear indication of when humans separated from other animals (because we are animals). As a result, we might be forced to accept that if we believe that humans have an immortal soul, then it could be argued that all animals that are a part of our evolutionary history must also have an immortal soul (and that evolutionary history will eventually take us back to single-celled organisms that lived billions of years ago).

Christian leaders generally teach that only human beings have immortal souls. The Catholic theologian Thomas Aquinas believed that all organisms have souls but only humans have immortal souls.[9] This belief also appears in Judaism.

On the other hand, in modern times there have been regular instances of the Pope, priests and pastors sometimes blessing animals. This is even known as an annual ceremony called the Blessing of the Animals (usually held during the first week of October). While Pope Francis suggested that animals have a place in heaven, this view is not held by people of more conservative beliefs. According to Italian news sources, the Pope said, "One day, we will see our animals in the eternity of Christ."[10]

If we were to accept the possibility that some, or all, non-human animals have an immortal soul, we could ask how this might affect our beliefs about eternal life. After all, other animals do not have Holy Scriptures to provide a plan for them to achieve everlasting life. If we're all in the same boat when it comes to eternity, why is it that humans require a guide (our Holy Scriptures) whereas other animals and our earliest ancestors (before the beginnings of formal religion) didn't require such a manual? For billions of years, life on this planet has existed without any Scriptures. What has caused this change within the last few thousand years?

What Does the Word Soul Mean?

It might be useful to consider where the word soul (especially as it relates to an immortal soul) came from.

For much of history, the soul was thought to be similar to our mind, that intangible ethereal part of us that was beyond our physical being. Rene Descartes, the famous French philosopher, called this a res cogitans (thinking substance) and res extensa (extended substance).

If we trace the concept of the soul back throughout recorded history we will begin to see that the word soul often referred to our mind or consciousness, or as some would describe it—our essence.

IS THERE LIFE AFTER DEATH? / 99

From earlier in this chapter, you might recall that the Egyptians believed that the soul was made up of various parts, although the heart (Jb) was believed to be the center of emotion, thought and will. Judgment in the afterlife for the Egyptians focused on the weighing of the heart. For the Egyptians, the soul was comprised of intangibles such as emotions, personality, thoughts and intent. Similarly, the Greeks believed that the soul (or psyche which might be described as life or spirit, although psyche could be literally translated as breath) was the seat of their mental abilities such as reason, emotions, consciousness, and thinking.

The Greek philosopher Plato, in *Phaedo*, presents Socrates' explanation of death: "Is it not the separation of the soul and body? And to be dead is the completion of this; when the soul exists in herself, and is released from the body and the body is released from the soul, what is this but death?"[11]

A brief look at what the Jewish, Christian, and Islamic religions believe about the soul will show us that the beliefs of ancient pagan religions are deeply rooted in the major religions of today.

Ministers and priests often quote Genesis 2:7 which reads that Adam "became a living soul" as proof that humans have souls. To do so is not actually correct. The Hebrew word in this verse that was translated into soul is *nefesh*, which means a living, breathing being, rather than an immortal soul. The correct interpretation of this verse should be that Adam became a living being.

In fact, the word *nefesh* is used in the Hebrew Torah to refer to animals also having this life force (Leviticus 17:11). To use this verse to support a belief in humans having a soul is incorrect, although as Judaism evolved, we begin to see a distinction between the body and the soul in their Scriptures. Most members of the Jewish religion believe that the soul was created by God in the beginning of time, somehow stored in a special soul vault, and that either at the moment of conception, or the moment of birth, the soul enters the physical body (earlier we looked at the same belief from the Egyptians). When a person dies, the opposite happens with the soul returning back to God.

The problem with this, as alluded to earlier in this chapter, is at what point in our evolutionary history did God decide to start taking souls out of this vault and implanting them into us? If you answer this question by stating that God did this when humans developed a greater intellect than other animals, then what happens to any humans such as the 8-year-old-girl, I mentioned at the beginning of this chapter, who possess limited intellectual ability?

If the receiving of an immortal soul from God is based on humans having greater intellectual ability than other animals, this would seem to be more than a little unfair. After all, more than any other creature, we have caused horrendous destruction to our Earth. One might ask if we really are the most intelligent creature on our planet?

If some animals have certain kinds of intelligence that is greater than that of humans, should they also have immortal souls, if the awarding of immortal souls is connected to intelligence?

Dolphins as an example have more structurally complex brains than humans.[12] Many animals are intellectually superior to children under the age of two. If you were to place a one-year-old cat alone in the wild and a one-year-old human alone in the wild, which of the two would be most likely to survive? Intelligence can be defined in different ways.

A young chimpanzee named Ayumu, involved in research at Kyoto University, was able to memorize the position of numbers on a screen much faster than any humans. The scientists studying Ayumu concluded that chimpanzees "simply have a better working memory for visual information than humans."[13]

If some animals possess certain kinds of intelligence that surpass humans, should they also have immortal souls? Or, is it simply possible that our awareness of having a soul is the result of our intellectual ability to believe in such a concept? If this is correct, then the concept of an immortal soul is a human invention that is nothing more than an illusion.

What Do the Major Religions Teach About Souls?

Let's look a little further into what the major religions of today teach about the soul because it is through these teachings that most people have formed their beliefs about immortal souls.

When a person dies, in Ecclesiastes 12:7 we read that the dust returns to the ground, but the ruah (the soul) returns to God. Further Jewish thought (and keep in mind like Christianity and Islam,

IS THERE LIFE AFTER DEATH? / 101

there are wide interpretations of scripture from ultra-conservatives to far-left liberals) teaches that someday God will judge the actions of the body and soul in partnership after returning the soul to the body at a resurrection.

Let's look next at Christianity regarding its beliefs in the soul, immortality and life after death. Christianity, similar to Judaism, also teaches that humans have a soul. Some Christian sects believe that the human soul enters each person upon either conception or birth which again is similar to some Jewish beliefs (and once again similar to Egyptian beliefs).

The main difference between Christianity and Judaism is that Christians believe that Jesus was the Son of God, and that salvation can be achieved through a belief in Jesus. Romans 10:9 reads, "If you declare with your mouth, 'Jesus is Lord,' and believe in your heart that God raised him from the dead, you will be saved." And, of course the most often quoted verse in this regard is John 3:16 - "For God so loved the world that he gave his one and only Son, that whoever believes in him shall not perish but have eternal life."

One of the first leaders of the Christian church to give us the Christian viewpoint on the soul was Origen (ca. 185-254) who wrote in the Origen De Principiis:

> *"The soul, having a substance and life of its own, shall after departure from the world, be rewarded according to it deserts, being destined to obtain either an inheritance of eternal life and blessedness, if its actions shall have procured this for it, or to be delivered up to eternal fire and punishments, if the guilt of its crimes shall have brought it down to this..."[14]*

Approximately 100 years later, Augustine wrote in *The City of God*, that the soul "is therefore called immortal, because in a sense, it does not cease to live and to feel; while the body is called mortal because it can be forsaken of all life and cannot by itself live at all. The death, then, of the soul, takes place when God forsakes it, as the death of the body when the soul forsakes it."[15] Centuries later, Thomas Aquinas taught that the soul is a conscious intellect and will, and cannot be destroyed.

The early Christian concept of the soul laid the foundation for similar beliefs within the Christian religion today although there are those today who tend to use the word spirit instead of the word soul.

Let's look next at Islam to better understand their concept of life after death. The Islamic religion had its roots in the early 7th century approximately 600 hundred years after the beginning of Christianity (although it could be argued that it began much earlier because some of the earliest people mentioned in Hebrew scripture are also considered to be prophets of Islam, such as Adam who they consider to be their first prophet).

In 610 CE, Muhammad began to receive what Muslims consider to be divine revelations. Similar to Judaism and Christianity, Muslims also believe in a day of resurrection for the dead accompanied by a final judgment (which includes either eternal heaven or hell). Muslims believe that Allah sends an angel to breathe a soul into the fetus while in the mother's womb. At death, the soul separates from the body.

Similar to Judaism, Muslims believe everything that God created has a soul, but only humans can experience immortality because we are the only creature who has free will (except that this creates a problem for any humans who don't have the intellect or ability to exercise this so-called free will), and this is also assuming that there are no other animals on our planet besides humans who have free will. If you believe that some animals are able to consciously make decisions, then the belief about humans having a soul because we have free will can no longer be accurate. Using the argument of free will, we would have to conclude that any animal that is capable of making a conscious decision must also have an immortal soul. And to take this a step further, if any animals other than humans have immortal souls, then belief in a religion cannot be a requirement to experience eternal life.

In the Islamic religion, similar to Christianity, death is the separation of the soul from the body and the beginning of an afterlife. Muslims believe that after burial, two terrifying angels appointed by God will question the dead to test their faith. The souls of righteous believers will then live in peace and comfort while the souls of sinners and disbelievers will experience punishments. Similar to Judaism and Christianity, in the Islamic faith there is a belief in a resurrection and final judgment at the end of the world.

The Judgment of Our Sins

In considering death, the Egyptians believed that life was a journey, that as the sun rose in the morning and set in the evening, each person traveled on a similar journey of birth, death and resurrection. The Egyptians believed in the continuation of life. They believed in an afterlife that had some similarities to life on earth (thus the reason for burying their dead with tools, food and in some cases even servants). They also believed that their placement in the afterlife depended on a judging of their life.

The judging—known as the Weighing of the Heart—was conducted by Osiris, the chief god of the dead and afterlife. The dead were expected to confess any wrongdoings to Osiris before he would determine their place in the afterlife.

Many beliefs surrounding life after death have been borrowed from the Egyptians by most of today's religions.

The Soul, Immortality, and Judgment

The Egyptians gave us three concepts that remain in the major world religions of today. These three concepts are the soul, immortality, and a judgment by a god (which includes the resurrection of the body and soul). We also find similar beliefs reflected in the religions of the Greeks and the Romans, civilizations that were all in place before the beginning of both Christianity and Islam.

The Greeks initially believed that the spirit of the dead would depart the body in a breath of wind and enter Hades, the Greek underworld. Homer describes Hades as a dismal place of wandering shadows.[16] When the dead were prepared for burial, a coin was placed on the tongue of the deceased so that they could pay Charon who was the ferryman to carry souls across the Styx river into Hades. Those buried without the coin were thought to wander forever along the shores of the river.[17]

Early Greek thinking about the afterlife was so dismal that eventually these beliefs evolved into something more positive. Some Greeks began to believe that they could be purified through mystic rites and through reverence for sacred texts. Others began to believe that those who lived a virtuous life would have a more pleasurable existence in the hereafter and there would be a more severe judging for those who did not please their gods while they were still alive.

A key Greek belief, similar to that of the Egyptians, was that there was a soul (the Greek word was psyche) that lived on after the death of the body. Accepted funeral traditions were considered to be pleasing to the gods and could help the soul to enter a more favorable form of eternal life. Even in wars, the Greeks requested permission from their enemies to retrieve their dead, so they could prepare the bodies for a proper burial or cremation.

Similar to the Greeks, the Romans had a plethora of gods, and similar to the Greeks there were also a wide range of cults that makes it difficult to identify one common religion and set of beliefs related to the afterlife, although there are some basic thoughts that underlie many Roman beliefs related to the afterlife.

The Romans believed in a gloomy place for the dead known as Hades. They also believed, similar to the Greeks, that the soul departed to the river Styx where Charon took the souls across the river, provided they paid him first. Once across the river, the Romans believed that the souls were judged by Minos, Rhadamynthus, and Aeacus, three demi-gods who were the sons of the god Zeus.[18]

What About Eternal Damnation?

Interestingly the Romans did not believe in eternal damnation nor did the Egyptians although the Egyptians did believe in a fiery hell, but not an eternal hell. Eternal hell would come later when it was perhaps necessary to frighten people to keep them from disobeying their religious leaders. Fear is a powerful motivator, and religions throughout history have used fear very effectively to control the masses.

I can clearly remember as a child being frightened by preachers who shouted with dramatic flair about unbelievers living in an eternal pit of fire. Could there possibly be a more powerful way to convince people to believe in a religion? Throughout the ages, could there possibly have been a more powerful way for church leaders to keep their followers in line like a flock of mindless sheep?

IS THERE LIFE AFTER DEATH? / 105

Funeral Practices from Pagan Religions
Judaism, which is the oldest of the three modern day major religions, is first mentioned in Greek records during the Hellenistic Period (323 BCE–31 BCE) although religion literature tells the story of the Israelites going back to around 1500 BCE.[19] Although Judaism began as a polytheistic religion (the belief in many gods) it evolved into a monotheistic religion (the belief in one God). Judaism wasn't actually the first religion to have a monotheistic religion; the Egyptian pharaoh Akhenaten founded the belief in one god before the Jews.

There are several scriptural verses that talk about the early Jewish people believing that there were a plethora of gods. Psalm 86:8 states, "There is none like you among the gods, O Lord." Psalm 135:5 reads, "Our Lord is above all gods." The very first commandment given to Moses states, "Thou shalt not worship any other gods before me" (Exodus 20:3). Over time we begin to see Judaism moving from polytheistic worship to monotheism.

Although Judaism began to evolve towards monotheism, and eventually both Christianity and Islam would have their roots in Judaism, the basic beliefs of these three major religions as they relate to life after death have their foundation in pagan religions. In the major world religions of today we see the concepts of immortality, the soul, and a judgment by God (including a resurrection) which all had their roots in the pagan religions of the Egyptians, Greeks and Romans.

Today, in most cultures, coffins (caskets) are used to house the dead. This practice is similar to what the Egyptians used (their coffins were known to them as sarcophagi). Like the ancient Egyptians, our caskets come with various designs and are constructed with a variety of materials, dependent on the financial means of the family of the dead person. An ancient Egyptian family might feel right at home selecting a casket in modern times. While wealthy Egyptians might have used a golden chariot to carry the coffin, we have replaced the horse-drawn vehicle with a glistening hearse (for those who can afford this practice). The preparation of a body for viewing at today's funeral home (which includes embalming) has its roots in the practices of the ancient Egyptians (although the oldest known deliberate mummification came from Chile in South America more than 7,000 years ago, which is about 2,000 years before this practice was used by the Egyptians).[20]

During a funeral service in the 21st century we would expect to hear some of the same terminology (such as heaven, soul, and resurrection) that was used by the ancient Egyptians (and was likely inherent in the beliefs of our ancestors as far back as 100,000 years ago).

Do Heaven and Hell Exist?

So far, we have seen the concept of an immortal soul threading its way through religions since the beginning of recorded history. Let's also take a look at heaven (and hell). After all, if we have an immortal soul it would be nice to know where it is going to be spending eternity.

Let's begin with Judaism. An important point related to Judaism and their belief of life after death is that the Torah does not reveal any details of what heaven is going to look like. Isaiah 64:3, in looking at the concept of heaven, reads, "Never has the ear heard it—no eye has seen it—other than God." If there is no specific description of heaven in the Torah (which provides the roots for Christianity and Islam), when people talk of heaven, where are they getting their information from?

The Old Testament (which as previously mentioned laid the foundation for Judaism, Christianity and Islam) teaches that death is similar to sleep (Ecclesiastes 9:5 and Psalms 146:4 and 115:17) although the New Testament teaches that believers will be immediately in the presence of Jesus (2 Corinthians 5:6-8; Philippians 1:23) while others might reside in hell (Revelation 1:18). For all, there will be an eventual resurrection where the body and the soul will be reunited, and God will judge each person, after which people will enjoy heaven for eternity or suffer in hell for eternity.

Purgatory, which is not directly mentioned in the Bible, is a Catholic belief that some people are not good enough to go immediately to heaven so they are placed in a "holding" area for period of time before they get to Heaven (although prayers and indulgences—which are really just donations to the church—are thought to reduce the time a person spends in purgatory). Purgatory still supports the belief of an immortal soul and the concept of heaven.

The imagery of heaven as written for Muslims in the Quran is more detailed than in the Christian and Hebrew religions. There are references in the Quran to heaven having beautiful gardens (Quran 55:54). The Quran also refers to heaven having beautiful women for men to enjoy (Quran 44:52-54) and it appears that these women are all virgins (Quran 55: 56-58). It should be noted though that some Muslim scholars believe that verses like this in the Quran are to be interpreted metaphorically, but the Islamic extremist who becomes a terrorist suicide bomber is proof that some Muslims interpret such verses literally (although interestingly suicide is considered to be a major sin and is prohibited in Islam, which is evidence that some people can be convinced—maybe deceived is a better word—to die for their religion, even when doing so is based on false teachings).[21]

In the major religions of the world, there are a wide variety of sects whose interpretation of scriptures vary, but there tends to be an underlying theme in the beliefs of life after death (the immortality of the soul) and a final judgment by God.

In looking at both polytheistic religions (from the Egyptians, Greeks, and Romans) as well as looking at monotheistic religions (such as Judaism, Christianity, and Islam) we find a similar core of beliefs related to the immortality of the soul.

What Do We Now Know About the Soul?

As our understanding of the human brain has increased in the 21st century, we have become more aware that our emotions, thinking, personality, and consciousness (the traditional words used to describe a soul) are rooted in the physical fabric of our brains. Is it possible that the concept of an immortal soul and heaven are false concepts that have been repeated so many times since the ancient Egyptians that they have become ingrained in us? Is it possible in the 21st century to look at eternal life in a different manner?

From a scientific standpoint we might ask whether it's possible that a soul could exist within us that is separate from our physical being. Physicist Sean M. Carroll has stated that for a soul to exist:

> "Not only is new physics required, but dramatically new physics. Within quantum field theory, there can't be a new collection of 'spirit particles' and 'spirit forces'

that interact with our regular atoms, because we would have detected them in existing experiments."[21]

We might very well ask why in the 21st century so many societies continue to bury their dead and support belief systems associated with death that began such a long time ago (and were borrowed from pagan religions by the major religions in our world today). One of the best answers to this question is that there is often comfort in following these traditions, especially considering that there has been such a long period of usage and acceptance. This leads to the obvious question, "Is there anything wrong with following these traditions?"

My answer to such a question is "no" providing that such traditions do not cause harm to anyone else, including the possibility of psychological harm to children. Bereavement can be very difficult. Any traditions that can help to ease the pain can be useful in helping a person to grieve which is an important part of dealing with loss. The loss of someone close to you can result in both emotional and physical suffering. If funeral traditions or beliefs in the afterlife help to provide comfort, then we should be willing to accept the unique approaches of different cultures in dealing with loss (once again stressing that such beliefs should not be harmful to others).

Robin Dunbar, an evolutionary psychologist, writes:

> *"It is surely no accident that almost every religion promises its adherents that they—and they alone—are the 'chosen of 'god', guaranteed salvation no matter what, assured that the almighty (or whatever form the gods take) will assist them through their current difficulties if the right rituals and prayers are performed. This undoubtedly introduces a profound sense of comfort in times of adversity."*[22]

Are Near-Death Experiences Real?

During the last few decades there have been a number of best-selling books that have attempted to provide a glimpse of life after death. These books are generally written by people who have had either near-death experiences or who were even clinically declared dead but revived. These experiences often include some of the following:

- *feeling weightless*
- *hovering above the body and looking down on it*
- *seeing the presence of a brilliant light at the end of a tunnel*
- *visiting hell and being tortured by demons.*

Some people, just before dying, report vivid dreams. A study, reported in the *American Journal of Hospice and Palliative Care* in 2014, found that such dreams "bring about a sense of peace, a change in perspective or an acceptance of death, suggesting that medical professionals should recognize dreams and visions as a positive part of the dying process."[23]

According to a Gallup poll reported in 2011, approximately 3% of the U.S. population have had near-death experiences.[24] Such experiences have been reported in a wide range of cultures with written records dating back to ancient Greece. While such experiences can be important in helping to provide comfort to a dying person (and to that person's family and friends), they don't in themselves provide actual evidence that heaven exists. In fact, some studies have shown that some people experience what appear to be near-death experiences even when they are not dying.

People who experience seizures have also reported out-of-body experiences.[25] In addition, Kevin Nelson, a neurophysiologist at the University of Kentucky, states, "At times of extreme danger or trauma, many people report out-of-the body experiences, seeing intense lights, or a feeling of peace."[26]

Neuroscientist Dean Mobbs, from the University of Cambridge's Medical Research Council Cognition and Brain Sciences Unit, states, "Many of the phenomena associated with near-death experiences can be biologically explained."[27] It has been found that out-of-the-body experiences can be induced by drugs or electrical stimulation of the brain, and that some people can even deliberately self-induce such a state.[28]

What Does Heaven Actually Look Like?

If near-death experiences cannot provide a credible view of heaven, is there some way we can determine what heaven actually looks like (if we accept that heaven actually exists, and there is absolutely no proof that it does exist)?

> *"A Pew research study in 2015 found that 72% of Americans believe in a heaven while 58% of Americans believe in hell. Focusing solely on Muslims, we find that 89% believe in heaven and 76% believe in hell. Less than half or fewer of Hindus, Buddhists and Jews believe in heaven, and approximately one-third of Hindus, Buddhists and Jews believe in hell."*[29]

While a significant number of American people appear to believe in heaven, we might ask what does heaven actually look like for these people? At funerals, it's common to hear people say things like "She's in a better place," "Now he's at peace," "She's looking down on us," or "He's probably up there playing golf." These are all typical comments frequently spoken at funerals as people provide their personal thoughts about the "passing" of a loved one.

In considering what heaven looks like, I Googled this question and found a Christian chat room where the discussion focused on answering this very question. Here were some of the thoughts (from Christian believers):

> - *"We will never run out of things to do, things to learn, skills to acquire."*
> - *"There will be dancing, singing and feasting in Heaven."*
> - *"There are animals and creatures in heaven."*
> - *"There will be all sorts of recreational things like sports, arts, all sorts of stuff to explore and do."*

For many in the 21st century, it is hoped that life after death will be a continuation of enjoying the pleasures (such as hobbies and interests) that they enjoyed here on earth, minus any pain or suffering. This type of thinking is more directly related to the beliefs of the Egyptians and our prehistoric ancestors than it is to the scriptures of Judaism and Christianity. Islam which is more recent than either Judaism or Christianity presents a more physical, pleasurable heaven (even if it caters to the needs of men more than women).

IS THERE LIFE AFTER DEATH? / 111

From my own experience, many of the informal conversations that occur at funerals have no basis in Holy Scriptures. As modern-day humans we have often fabricated our own interpretation of heaven to help us better cope with the death of a loved one, and/or to create personal hope for our own life after death.

The major religions of today have streamlined and formalized the basic beliefs of ancient civilizations such as the Egyptians as they relate to the soul, immortality, and a final judgment, but individuals often insert their own hopes and interpretations of heaven into interpreting these concepts. While we are moving in the 21st century towards more humane thinking about the afterlife, let's not forget that there are some who believe that the afterlife will be the ultimate form of vengeance when horrific punishments will be unleashed on unbelievers. The influential Christian theologian and philosopher Thomas Aquinas, in talking about heaven, said that in addition to contemplating God, the greatest pleasure in the afterlife would be watching the eternally damned being tortured.[30] If any religion has compassion and love as part of its teachings, how could any member of that religion find satisfaction in the belief that non-believers will suffer eternal punishment in a fiery hell? The concept of hell is a contradiction to the belief that God is love.

Heaven and Natural Selection

It would appear that any scientific laws that hold true here on earth also maintain their validity throughout the universe. While many people view heaven as a place of eternal bliss, perhaps pursuing one's hobbies or bathing in the glory of the angels singing and playing their harps, such a picture is a contradiction to the laws of science.

The law of natural selection that drives change here on earth also appears to drive change throughout the universe. If God exists, he cannot simply discard the basic laws that are uniform throughout our universe to create heaven. If heaven exists, it must also follow the basic scientific laws that provide the foundation for our universe.

As a result, if heaven exists, it can't be a static place. It must continue to follow the law of natural selection which would also apply to the spiritual beings of heaven. The constant struggle that we experience here on earth will not end in heaven. Survival of the fittest is a basic law that impacts all life within our universe. Even if heaven is somehow in a dimension that we haven't yet discovered, for it to be a static place where change does not occur would be a contradiction to creation itself.

As we have already seen in this book, natural selection is a universal law that drives change. If heaven exists, it must also abide by this law. Natural selection would dictate that the struggle to survive that we face here on earth would continue in heaven.

Ancient Guides for Living

We all use the most up-to-date 21st century technology in every aspect of our daily lives, including computers, the Internet, transportation, education, entertainment, and the list could go on and on.

Why is it that 21st century advances impact almost every area of our lives, with the exception of religion which has such a direct impact on our search for meaning? Why is it that so many people cling to ancient religious texts to find personal meaning when some of the content within these so-called Holy Scriptures is often used to promote war, misogyny, intolerance, racial discrimination, persecution of non-believers, guilt, fear, and anxiety?

In Chapter 7, we will look at how 21st century learning is beginning to reshape our thoughts about immortality and heaven. Before we do this, let's take a look in the next chapter at why it is so difficult for many people to embrace change. And more specifically, why is it so difficult for the conservatives in any major world religion to accept new learning? After all, it is the religious conservatives who cannot accept climate change or would rather have a major war to defend their faith than reach out and accept the faith of others.

- 6 -
Why Do Some People Resist Change?

Nobel prize winner Bob Dylan wrote a song titled *The times they are a changin'*.

The times have always been changing. In previous chapters we saw how evolution, as driven by natural selection, has been relentless in causing change to every lifeform since the inception of our planet. The major difference in the 21st century from previous centuries is the rate of change that we are experiencing as technology explodes around us. In these fast-changing times there are some people who are ferociously clinging to the traditions and beliefs of the past while others are challenging these long-held beliefs, and there are still others who are racing ahead to forge new ways of thinking with little regard for the past.

In the first five chapters of this book we looked primarily at how the past has shaped our current thinking, especially in response to the questions of why do pain and suffering exist and is there eternal life. Before moving on to explore how science and technology may be creating new answers to these questions, let's take a look at why some people are reluctant to embrace change.

In my twenties, I attended the University of Toronto to complete my Master's Degree. My area of specialization was in counseling. As part of this program, one day each week a group of six of us practiced our counseling skills in real-life situations with a registered psychologist supervising us.

On one occasion in our counseling practicum, our client was a woman who was in an unhappy marriage, and as a result wished to leave her marriage. The supervising psychologist chose me to work with this woman.

After counseling the woman for about forty-five minutes, I felt good about what I had accomplished.

My instructor then asked the client what she thought of my efforts to help her. She responded that she felt I had tried to force my values on her without really listening to her concerns, that I kept trying to find ways to save her marriage when she really wanted to get out of it.

As the other students in our class also evaluated what I had done, I heard the same criticism, except from one of my peers who happened to be a member of an evangelical church. This female evangelical classmate felt I had done an excellent job in trying to help this woman save her marriage. For both this classmate and myself, our approach to helping this client was influenced by our shared beliefs, as taught by our churches, that marriages were sacred.

The psychologist in charge of our class was a little more forceful in his comments to me (and my evangelical peer). As my instructor bluntly said, "Counseling is about helping a person resolve his or her concerns without imposing your values on the process." In addition, he explained that I had made no attempt to discover the actual facts of what was happening in the woman's marriage. I had blindly accepted whatever she said without making any attempt to establish the credibility of what I was hearing.

These were important lessons for me to learn as a beginning counselor, but these comments also had implications for nurturing the seeds of doubt in my mind concerning my religious upbringing. Up until this point in my life, my values had been dictated by others (primarily my church in its interpretation of the Bible). The way I lived my life was largely the result of obedience to the teachings of a church, not the result of my own free will (and this is not to say that some of these values were not positive).

Challenging Our Beliefs

Although I had growing doubts about some of the teachings of my church as I grew up, I never really challenged them. At one point in my early twenties, I spent a significant amount of time reading and studying the Bible from cover to cover. In doing this, my interpretation of what I was reading was shaped by the teachings of my church as well as the conservative authors of Bible guides that I used to help me. I don't ever remember any of these authors pointing out any concerns or discrepancies that might exist in the Bible. I had been taught to accept the interpretation of any content of the Bible by those who were considered to be the authorities, and that's exactly what I did. I allowed the values of others to shape my values (without knowing all the facts), and this had been occurring to me from very early in my childhood.

I had completed two university degrees with an underlying emphasis on critical thinking, and yet I failed to apply these critical thinking skills to my own system of beliefs.

Blind Obedience or Fact Checking

During the following few years as I began a career as a counselor, I became more and more aware of how some of the problems that people experience are often rooted in their blind acceptance of the expectations that other people (and institutions such as churches) have for them. In addition, I found that many people suffer personal problems as a result of irrational thinking, or from harmful thinking that results from mental exaggerations without having considered all the facts. As I became more skilled in helping others, I found myself beginning to strip away the blind obedience to the teachings of a church that had literally controlled my life.

I was 30 before I was able to step completely away from the church. Given that many of my friends also belonged to this church, I maintained contact with them for a while but soon it became apparent that they were uncomfortable in my presence, and I tired of their constant attempts to bring me back into the flock, although decades later I find that most of them have also left the church.

Leaving a church and my friends who belonged to the church was a huge decision, one that initially resulted in the loss of lifelong friends, as well as one that made it difficult for a time to be with my parents (who would eventually respect my decision).

The Comfort of the Past

Letting go of the past is not easy and it may be exponentially harder if you grew up in an environment that had strictly defined beliefs about religion (or some other bias related to the meaning of life).

In a 2007 poll reported in *Time Magazine*, people were asked how they would respond if science disproved one of their religious beliefs. Sixty-four per cent of respondents said they would reject the findings of science in favor of maintaining any of their religious beliefs that science had proved wrong.[1] Such thinking in the 21st century is unbelievable.

If we were to ask any of these people whether they enjoy modern scientific advances such as computers, the Internet, cars, and so on, I think we know what the answer would be. How is it that some people still ferociously cling to beliefs from thousands of years ago in the area of religion while these same people enjoy all the latest advances that technology brings?

As we journey through life, it is often more comfortable for most people to accept what they were taught when they were growing up then it is to shift gears and go in a new direction. Some people allow past beliefs, without every questioning them, to dictate how they will live their entire lives, even if they remain unhappy.

The Return of Christ

By the time I was in my early twenties, I was conducting Bible study groups for both teenagers and adults at the evangelical church that I attended. At this time, there were some popular, yet controversial, books on the market suggesting that we were living in the end times, a theme if one were to study the actual history of Christianity that has been with us constantly for the past 2,000 years. The earliest believers of Christ expected his return at any moment, and 2,000 years later Christians are still waiting.

Getting back to those trendy books about the coming apocalypse, I remember presenting some of the controversial thoughts from these books in my Bible study groups. Although there were occasional questions, no one challenged anything I said. Without any formal Bible training, I had become the de facto expert in my church, and like the ordained minister, no one questioned anything I said. In retrospect, the books I was referring to in my Bible study groups were about as credible as the trashy magazines that sit on the racks at the checkout counters of our grocery stores.

Blind Acceptance

For many hundreds of years, the Catholic church kept the mass-

WHY DO SOME PEOPLE RESIST CHANGE? / 117

es ignorant of the contents of the Bible. Only priests could read what was found in the Bible and you can bet that in their interpretation of what they were reading there was a great emphasis on obeying God. You can also bet that the priests made sure that their members were well aware that the priests were God's representatives here on earth and as such their commands (and desires) were indisputable.

For hundreds and hundreds of years, the culture of the church has been one of the masses blindly accepting whatever a church leader said. Questioning the words of a priest, rabbi, or minister was out of the question. The very organizational framework of almost all church services is even today one that prevents the congregation from asking questions or from refuting what a church leader is saying. Blind obedience is the hallmark of most religions.

In the Bible study groups that I taught at my church, I was astounded by the lack of understanding of the Bible by many adults. Some of the adults in my Bible study groups were many decades older than me, but other than their favorite verses (John 3:16) or stories of Adam and Eve, or Noah and the ark (neither story which has any historical support), for the most part they had minimal awareness of the contents of the Bible. Supporting my observations, a 2012 Pew Research survey found that the group of people who answered the most correct questions about the Bible were the atheists.[2]

During the same period of time that I was teaching Bible study groups in my church, I was also teaching a history course in a local high school. I was often amazed that my high students had endless questions about the content of what I was teaching. They often engaged in heated discussions. But at my church, in any Bible study groups that I taught, there were few questions and there was most certainly never any disagreement. From a very early age, members of evangelical churches (of any conservative religion) are often taught to never question authority, and remarkably without any Bible training I had become one of those authority figures in my church.

I was also involved in a Bible study group with college and university educated people where there was a more open discussion, but even in this setting the tendency was to dance around any verses that might cast the contents of the Bible in a negative light by using phrases such as, "God works in mysterious way," or "Sometimes his truth is beyond our understanding." In other words, we abandoned our intellect and blindly followed the teachings and traditions that had been passed on to us. There was never a real effort to get at the truth.

Christians are not alone in failing to study the contents of their Bible. Millions of Muslims, while placing total obedience in their sacred writings, have little understanding of these actual writings. Many Muslims have been taught to read and speak Arabic (the language of their Holy Scriptures), although a significant number of them do not understand what they are actually reading or speaking.

Waqar Rasool, a Muslim who works at the Change Initiative, writes:

> *"I was born into a Muslim family. And as a kid was taught the basic concept of Islam. An Imam of the local mosque was 'hired' by my parents to come to our home to teach me and my sister how to read Arabic. Starting from the age of 4 or 5. By age 12, under guidance of different Imams over time, I had read the entire Quran. Although I had finished reading the Quran by age 12, I had no idea of what I had read. ABSOLUTELY no idea."*[3]

The Quran and the Bible are the two leading books of Holy Scriptures, yet millions (perhaps even billions) of Muslims and Christians have little knowledge of the extensive contents of these books, other than a handful of verses that are regularly chosen by their leaders.

Living outside the U.S., as I do, it's common to meet people who wonder in disbelief how Americans could have ever chosen Donald Trump to be President of the United States, given that he is racist, misogynistic, homophobic, Islamophobic, and even brags about sexually assaulting women. Even more puzzling is how the religious right (evangelical Christians) who claim to be the moral conscience of the country and who wield significant voting power in the U.S. could support such a man, a person who appears to be the antithesis of their beliefs. It seems that whatever Trump says or does, Christian conservatives in the U.S. never waiver in their sup-

WHY DO SOME PEOPLE RESIST CHANGE? / 119

port him. It's almost as if someone has taken their intellect and conscience away. How did this happen?

Let's look at how conservative Christians could be so blind, and in the process perhaps you might better understand why some people are unable to accept change.

As I have previously mentioned, I grew up in an evangelical church. I won't name the actual church because what I experienced is not much different than what others experience growing up in any home that is permeated by a conservative religion. I would like readers to think in a broad sense in looking at what I am going to say, rather than pigeon-holing a specific church. And please don't get me wrong: I had very loving parents who I'm sure never fully understood how an evangelical upbringing really affected a person.

When I was a child, probably less than a year old, I was involved in a ceremony in front of the congregation of my church. During this ceremony (called a dedication ceremony), my parents made a public vow that they would raise me as a child to love and serve God. According to this church's website, the purpose of the dedication ceremony was the "setting apart of the children to be the servants and Soldiers of the Lord Jesus Christ." Before I was one year old, I was dedicated by my parents and church to be a Soldier of Christ.

While all fundamentalist churches may not have exactly the same dedication ceremony for babies, the intent of fundamentalist churches is the same—raise their children to serve God and attempt to convert others (and subscribe to the beliefs of this particular religion without ever challenging it). James Dobson, the founder of Focus on the Family and an evangelical Christian, whose voice is carried on more than 2,000 radio stations across the U.S., said:

"Those who control what young people are taught, and what they experience— what they see, hear, think, and believe—will determine the future course for the nation."[4]

During the course of a normal week when I was growing up, I often attended church two to three times on Sunday. In addition, to give you an idea of my own family: my father attended church related meetings on Monday nights, my mother attended choir practice on Tuesday nights, I attended junior band practice on Wednesday nights while my parents attended Bible study. On Thursday nights my father attended senior band practice and on Friday nights I attended youth group activities. Although Saturday tended to be a free night, there were a fair share of musical or other activities on Saturday nights along with a married couples' group for adults.

During all of this church involvement, the thought of ever questioning anything related to what was being taught was taboo.

Kimberly Blaker writes, in her book *The Fundamentals of Extremism*:

> *"Fundamentalists know too well that children who learn to think on their own may someday stray away from their indoctrination. The ideology of children in fundamentalist families is predetermined. Mind control, therefore, is the mode by which fundamentalists, whether Christian, Islamic, Jewish, or any other group gains adherents. If fundamentalists do not guard against children learning to think on their own, they risk turning out adults who will choose a path inharmonious or even opposed to their own."*[5]

For me, the indoctrination of our church was all-encompassing. Even before I started school, I attended Sunday School and then I sat through church service after church service, subconsciously absorbing the doctrines and traditions of the church. I would estimate that between the age of 2 and 16, I probably attended over 2,000 church-related services, and the message was mostly the same—"Repent your sins and be saved, otherwise you will suffer in a fiery hell for eternity." And the emphasis was primarily on avoiding the fiery hell rather than the looking forward to the rewards of heaven.

Fear is a powerful tool in the hands of fundamentalist leaders. The day will come, if it's not here already, when people begin to realize that frightening young children into believing any religious doctrine is a form of abuse.

I would like to add that in spite of my rigorous church upbringing, I lived in a loving family where my parents instilled many positive values such as caring for others, learning a positive work ethic, and gaining a love of music. I don't believe my parents consciously thought that their church was harming any of their children. They had grown up in the church at a time when blind obedience to authority was a way of life. Times have changed, but unfortunately what I was exposed to at church has continued for many children who are taught never to question what they are hearing.

Although more and more young people are leaving the church of their parents today, unfortunately many of them are leaving with the teachings of the church stuck in their minds because of their early indoctrination.

Between the age of 7 and 14, I was known as a junior member (actually a "Junior Soldier") of the congregation at my church. Once a year, either in Sunday School or in front of the main church congregation, Junior Soldiers would sign a commitment which read:

> *"I know that Jesus is my Savior from sin. I have asked Him to forgive my sins, and I will trust him to keep me good. By His help, I will be a loving and obedient child, and will help others to follow Him. I promise to pray, to read my Bible, and by His help to lead a life that is clean in thought, word and deed. I will not use anything that may injure my body or my mind, including harmful drugs, alcohol or tobacco."*

Now keep in mind that I began to publicly sign such a document when I was only seven years old. Staggering, isn't it? Recently I was looking through some old photo albums and I found a copy of this declaration that I signed when I was 8 years old. In reading it now as an adult, I was appalled by the words "obedient" and "helping others to follow Him."

The word "obedient" underscores that young people raised in a conservative religion are never expected to question anything to do with the church or Bible, or question their parents (more on this later when we look at the abuse statistics in fundamentalist homes).

The words "helping others to follow Him" also found in the declaration that I signed as a seven-year-old suggests that this religion considered itself to be superior to others. Instead of there being something in this document about accepting the religious beliefs of others, from a very early age I was forced to accept one way of thinking, that somehow Christianity was better than other religions.

There was nothing in this declaration about being kind to others or being the best person I could be. The declaration was all about serving God (or really, the church) obediently and growing up with a mindset that my mission was to convert others to Jesus.

At the age of 14, in my evangelical church, junior members become senior members ("Senior Soldiers"). Once again, I was expected to sign a document in front of the congregation pledging my obedience. Before signing the declaration, along with other teenagers, I attended classes where I learned more about the doctrines of my particular church, doctrines that focused on beliefs such as:

> - *We believe that the Scriptures of the Old and New Testaments were given by the inspiration of God, and that they only constitute the Divine rule of Christian faith and practice.*

> - *We believe that our first parents were created in a state of innocence, but by their disobedience, they lost their purity and happiness, and that in consequence of their fall, all men have become sinners, totally depraved, and as such are exposed to the wrath of God.*

> - *We believe in the immortality of the soul, the resurrection of the body, in the general judgment at the end of the world, in the eternal happiness of the righteous, and in the endless punishment of the wicked.*

So, at the age of 14, I signed this declaration in front of the congregation (after all, I had been taught to be obedient; not signing the document was not a choice). Even though I knew very well that Adam and Eve were not our first parents, and that their story, at best, was mythology, I signed the document. After all, if I didn't sign it, I would suffer "in the endless punishment of the wicked."

While I'm not going to provide every detail of the declaration that I had to sign, here are a few phrases from the declaration:

- *"endeavoring to win others to Him"*

- *"loyal to its (the church) leaders"*

- *"giving as large a proportion of my income as possible to support the church"*

- *"I now call upon all present to witness that I enter into this covenant and sign this declaration of my own free will."*

At the age of 14, I would hardly say that I signed the document out of my own free will. At any rate, I attempted to live up to what I had signed, although throughout my teenage years I often had a knot in my stomach that something was not quite right with the beliefs I had subscribed to. And then there's the word "loyal" again in the declaration, although this time a financial component had also been added.

By my mid-twenties I was now a youth leader in my church and even taught adult Bible study groups. This might have continued forever had I not had agreed one summer to be the director of a summer camp that was operated by my church for "underprivileged children."

Children from a large city were chosen (from families that various social agencies worked with) to attend camp for a week. I think we had between 6-8 sessions over the course of the summer (and probably over 100 kids during each session). For anyone who has ever worked at camp or has been a camper, you are likely aware that there is generally a quiet time after lunch and there is often a storytime of sorts before bed.

During these quiet times, I quickly discovered that my staff (mostly senior high school students and college students) were reading their Bibles and praying with these young children (some as young as 8) in an attempt to convert them (keep in mind that most of these children did not have a church background). Suddenly I saw what I had been blinded to for much of my life. Forcing one's religion on other people is not appropriate, especially if one is trying to convert children (and in a situation where they have been removed from their parents and are essentially trapped in a camp several hours from home). And one could also argue that if the main reason for a church, or any of its members who are engaged in any form of charity work, is to convert others to their religion, then their charity motives are questionable.

By the time the summer ended, I was disgusted to be a part of a religion whose main purpose was to convert others and that believed it had the only "true" answers for finding meaning in life.

Although it took me another few years to actually leave the church, I felt a tremendous relief when I was finally able to let go. Since that time, I have grown more and more cognizant of the harm that evangelical or conservative churches do by robbing people of their free will and coercing people to be obedient.

At the time of writing this book, the evangelical Christians in the United States were instrumental in electing an American president who is sexist, misogynist, racist, narcistic, out-of-touch with the needs of other people, and actually completely unaware of the basic beliefs of the evangelical Christians who voted for him. As Bryan Mealer writes in the *New Republic*:

> *"It's hard to decide which truth actually stings worse, that white evangelicals sold out Christian values for a couple of seats on the Supreme Court, or the grim prospect that our rigid Christian upbringing with all its trauma and guilt was nothing but a lie; it was never about the Good News at all, but white nationalism maintaining power through slavery and Jim Crow and now against a color-shifting, globalized society. When Donald Trump ascended to the White House, evangelical Christians suspended their moral convictions and followed him like a dime-store Messiah."*[6]

You might ask why evangelical Christians would vote for someone who is the antithesis of their values. Here are a few thoughts.

First of all, as previously outlined in this chapter, evangelical Christians (similar to many other fundamentalist religions) indoctri-

WHY DO SOME PEOPLE RESIST CHANGE? / 125

nate their members into blind loyalty from an early age. Disobedience is frowned upon. Evangelical Christians are taught to believe in an infallible Bible, but even more to the theme of this chapter, they are taught to blindly follow their leaders (whose job it is to interpret the Bible according to the church's needs). And when their leaders said to vote for Trump, like a flock of mindless sheep, their members followed.

For any readers who feel trapped by past beliefs that have been forced on you (even if these beliefs did not result from being involved in a religion), I understand what some of you are going through. For other readers who may have had the luxury of growing up in a home where you were free to choose your own beliefs, hopefully you can understand how some people have a great deal of difficulty letting go of past beliefs even when some of these beliefs may be built on false facts. As Jerry Coyne writes in his book *Faith vs. Fact*:

> "...*the main reason people turn a blind eye toward implausible beliefs is that they get their faith not through reason or deliberation, but through indoctrination from their family and friends.*"[7]

In the remainder of this book we are going to look at how new advances in the 21st century might impact our search for meaning. You may agree with some of what is presented, or you may disagree, but in order to fully appreciate how technology and science in the 21st century are going to impact our beliefs, there is a need for all of us to at least approach such a dialogue with an open mind.

Lobotomy?

If you (or someone else in your family) suffers from mental illness, would you like someone to drill into your head (with no anesthetic) to release the demons in your head as was once done, or would you prefer to be treated by a 21st century medical professional who has all the most up-to-date treatment strategies to help you?

If you wanted to have a house built for you, would you prefer to employ a 21st century builder who would use the best materials and environmentally friendly strategies, or would you prefer to have someone from the distant past build you a mud house with a thatched roof?

If you wanted the very best education for yourself or your children, would you turn to a 21st century school that was employing the best proven teaching techniques along with state-of-the-art technological equipment, or would you prefer to receive your education where the focus was on memorizing facts from outdated textbooks that had no longer had any application to our modern world?

Some people want to enjoy the benefits of the latest technological and scientific advances, yet when it comes to answering the question of finding meaning, they turn away from new learning and satisfy themselves with what may be ancient myths and false facts.

Churches Are A Changin'

In the previous chapters we looked at the law of natural selection. Natural selection does not only apply to biological development; it also impacts religion and philosophy. At a time when more and more people are turning away from organized religions, new religions will appear that better meet the needs of people, or old religions will evolve to survive, or unfortunately some people—unable to deal with change—will tenaciously grasp fundamentalist religions.

For many centuries the Catholic Church resisted change. In the 21st century as more members turn away from the Catholic Church (or any other church), the church will be forced to change (and this is already happening). Such changes have little to do with new revelations from God, and everything to do with the costs of maintaining the physical structures and administrative network of the religion. The Catholic Church, like many other churches, will change because the bottom line is that they need money to survive. Plain and simple, less members equals less money.

On the other hand, as long as conservative religions maintain their hold over enough members to meet their financial obligations, these churches will resist changing, but I guarantee that the minute that members begin to leave in significant numbers, and that particular religious denomination can no longer afford to pay for their buildings and ministers, doctrines will change overnight.

In the evangelical church where I grew up and often attended church three times on Sunday, the same church now only has one service on Sunday. It's an old adage that if you want to be a successful salesperson you have to understand the needs of your customers. As more and more people escape from the tyranny of conservative religions, these churches will change their ways to survive—it's the law of natural selection!

I'm not just talking about fundamentalist Christianity either. In this morning's paper I read that half of Iranians say "no" to veil law. Apparently a three-year-old report shows that nearly half of Iranians (both men and women) want an end to the requirement that women cover their heads in public.[8] The times, they are a changin'.

Continuing along with this thought, let's consider some recent changes in the Catholic Church. For instance, in Africa (which has had fastest growing influx of members over the last 100 years) the Catholic Church has accepted a blend of native and Christian expressions.[9] Compare this to the narrative of the Catholic Church who throughout hundreds of years of history often slaughtered anyone who didn't completely renounce their Indigenous religions and accept only the teachings of the Catholic Church (such as the slaughter of Indigenous people of the New World).

Sanitizing the Bible

To better meet the needs and beliefs of some of their members, churches will also sanitize passages in the Bible that may not be acceptable in the 21st century.

For example, the Bible teaches, "Whoever spares the rod hates their children, but the one who loves their children is careful to discipline them" (Proverbs 13:24, New International Version), while in the God's Word Translation this verse reads, "Whoever refuses to spank his son hates him, but whoever loves his son disciplines him from early on." This verse is particularly interesting because in another translation (the International Standard Version), the verse reads, "Whoever does not discipline his son hates him, but whoever loves him is diligent to correct." In this translation, God's command to strike a child with a rod (or some other object) is gone. This is a perfect example of how religions are beginning to change some of God's commands as presented in past versions of the Bible.

How can any Holy Scripture be the infallible word of God if

people are free to alter the meaning of verses in different translations? We might well ask to what degree did Biblical authors in the beginning years of Christianity (or any other religion) change the content of the scriptures to support their agenda and the culture that existed at that time.

Good Old Christian Values

Although some churches will evolve new ways of thinking in the 21st century and beyond, other churches (primarily fundamentalist churches regardless of their specific religion) will fiercely cling to the past. The fastest growing church affiliation in the U.S. is fundamentalist Christianity who preach a return to the "good old" values (although as mentioned in an earlier chapter, even evangelical churches in the U.S. are losing their younger members in massive numbers). If one were to explore these values further, we would find a male dominated authoritarian approach that had and still has more than its share of problems (which is perhaps one reason why the Christian Right in the U.S. finds it so easy to support someone like Donald Trump). Higher divorce rates, a greater prevalence of sexual and physical abuse, are all related to the good old values of conservative evangelical religion.

Consider the following examples of the good old Christian values. Sexual abuse by Catholic priests has been well documented and publicized, but what about abuse in the Protestant church? In 2015, the Methodist Church of Britain apologized for failing to protect children and adults, following a report that uncovered 1,885 incidents of abuse in the Church dating back to the 1950s. Between 2008 and 2011, Islamic religious schools in Britain faced more than 400 allegations of physical abuse.[10] In 2007, the three leading U.S. insurance companies said that they received an average of 260 reports per year of child sexual abuse at the hands of Protestant church leaders and members. Comparing this with 228 credible accusations within the Catholic Church during the same time period, we find that abuse in the Protestant church, according to insurance claims, is actually greater than that of the Catholic Church.[11]

In Queensland, Australia, Dr. Lynne Baker cites a study that found that 22% of perpetrators of domestic violence and abuse go to church regularly.[12] Boz Tchividjian, a university law professor who also runs an organization to assist abuse victims (and he is also the grandson of Billy Graham, the famous evangelical preacher) states, "The Christian mission field is a magnet for sexual abusers."[13]

James Dobson, the founder of Focus on the Family (an evangelical organization) states, "Spanking can be a valuable tool—if it is administered appropriately," and on Dobson's website, he outlines five reasons why spanking might fail. One of those five reasons is stated as "The spanking may be too gentle. If it doesn't hurt, it isn't worth avoiding next time. While being careful not to go too far, you should ensure he feels the message."[14]

A 1997 Canadian study, the largest of its kind, would disagree with Dobson. The study found that adults who were spanked or hit as children were twice as likely to suffer from an anxiety disorder as their peers who were never spanked, as well as experiencing higher rates of drug and alcohol abuse and depression.[15]

So much for good old Christian values.

While the U.S. is considered to be the most religious of all industrialized nations, it has a murder rate which significantly surpasses all other industrialized countries. And Japan, with a very low percentage of Christians, has the lowest violent crime rate of any industrialized country in the world. One U.S. state—Louisiana—has the highest church going rate in the country, yet it has twice the national average of murders.[16]

Religion and the State

Clinging to past beliefs will create tension and likely even violence as some people will feel threatened by those who seek new forms of meaning in the 21st century. Some people will blindly attempt to maintain some disturbing beliefs about discipline, homosexuality, sex, and woman's rights, although they may continue to express Biblical teachings about loving your neighbor.

The American fundamentalist Christian base also believes that their brand of Christianity should provide the foundation for any decisions made by the U.S. government (even though the Founding Fathers of the country were clear that state and religion should be separate). In fact, there is no mention of God in the American Constitution, even though every president feels compelled to end every speech with the words "God bless America."

Christians at War

Unfortunately, even today, many people who belong to conservative religions see themselves as being at war with the rest of the world. Consider these words from Randall Terry, the founder of the anti-abortion organization Operation Rescue:

> *"I want you to just let a wave of intolerance wash over you. I want you to let a wave of hatred wash over you. Yes, hate is good...our goal is to be a Christian nation. We have a Biblical duty: we are called by God, to conquer this country. We don't want equal time. We don't want pluralism."*[17]

As I will outline in the coming chapters, the 21st century will bring incredible opportunities to solve poverty, hunger, disease, oppression and wars, but unfortunately the very groups that preach love will likely be those who resist these changes until they see the opportunity that some of the technological and scientific advances might present to them in their mission to force the rest of the world to accept their beliefs. And unfortunately, the very unrest that 21st century technology will create will lead some people, at least initially, to fundamentalist religions. As Richard Swift writes in the New Internationalist:

> *"The causes of fundamentalism, while complex, are most often seen as rooted in the failed promise of modernity. This can be understood as human progress that has taken place in recent centuries. The extreme change of pace, especially during this century, has left people with constant pressure to adapt their habits and beliefs. All of this confusion has created the perfect environment for fundamentalism to breed."*[18]

Where Are We Going Next?

Let's move on to the next chapter and begin to look at where science and technology are taking us in the 21st century, and then later in this book (Chapter 11) we will take a look at how we can find meaning in the 21st century.

- 7 -
Will We be Able to Create Eternal Life in the 21st Century?

As we explored in Chapter 5, the concept of life after death—which appears to have been an important idea since the beginning of humankind—has largely been shaped by ancient pagan religions in civilizations such as the Egyptians. The major religions of today (Judaism, Christianity, and Islam) have borrowed heavily from ancient religions in giving us our modern beliefs about eternal life. We might ask, "How is this going to change in the 21st century?" Consider the words of Michio Kaku, a renowned physicist and professor of Theoretical Studies at the City University of New York:

> *"By 2100, our destiny is to become like the gods we once worshipped and feared. But our tools will not be magic wands and potions, but the science of computers, nanotechnology, artificial intelligence, biotechnology, and most of all, quantum theory, which is the foundation of the previous technologies....we will create perfect bodies, and extend our life spans."*[1]

As we also looked at in Chapter 5, the major world religions often focus on three basic concepts in looking at eternal life: the immortality of the soul, some form of judgment by God after death, and an eventual resurrection and reunification of the soul and body. We also considered how many individuals tend to have their own concept of what heaven looks like. For the most part, individuals hope for some continuation of the pleasures they enjoyed on earth while eliminating the pain they might have suffered. Finally, we considered how natural selection would impact heaven, if heaven actually exists. Natural selection, if it applies to all of creation (which it appears to), would tell us that if heaven exists any type of life there must continue to evolve, and that change (and struggles) would be a constant just as it is here on earth.

The major religions of today (having evolved from the major religions of our distant past) have given us the hope of eternal life. That hope is largely the result of the fact that our ancestors had a very short lifespan and a life full of daily hardships to avoid constant pain and suffering. For much of human history, it was rare for people to live past their 20s. It's no wonder that our ancestors needed to develop a belief in a heaven. In the 21st century as we begin to dramatically increase the length of our lives, enjoy better health, and even develop the potential to create our own immortality (with a personal heaven exactly the way each individual would like it to be), how will this impact the major religions of today that are largely based on the needs and culture of people who lived thousands of years ago?

Will We Be Able to Stop Aging?

It is likely that scientific and medical advances in the 21st century will significantly prolong our lives. Vladimir Skulachev, the Dean of the School of Bioengineering and Bioinformatics at Moscow State University states:

> *"Given that aging is programmed, it can be stopped like any other program. In that case, ailments characteristic of old age will be nipped in the bud; we won't age."*[2]

Our Increasing Life Expectancy

Earlier in this book it was noted that in the past thousand years the life expectancy of women has more than tripled from 25 years to 80 years in more than 50 countries in the world. Let's look at a shorter period of time. Since 1900, the global average life expectancy has more than doubled, now approaching 70 years worldwide and 80 in many countries. No country in the world currently has a lower life expectancy than the countries with the highest life expectancy had in 1800.[3] The UN estimates that the proportion of people over 60 will double before 2050.[4]

In the 1960s biologists estimated that the average threshold of human life was 89 years. It was predicted that even with improved health systems that there was a fixed limit as to how long we could live, just as there appears to be for other animals on our planet. The prediction was that we would live healthier longer, but there would still be an average fixed expiration date on life itself. Since these predictions in the 1960s human life span has started to increase markedly, going beyond what experts had once expected.[5] Today, biologists believe that the average life span has increased to 97 years, a gain of 8 years in a relatively short amount of time. Dr. Aubrey de Grey, chairman and chief science officer of the Methuselah Foundation, writes in his book *Ending Aging*:

> *"We can probably eliminate aging as a cause of death this century—and possibly within a few decades, soon enough to benefit most people currently alive."*[6]

While some experts believe that a scientist like de Grey is far too optimistic in his outlook, we should consider the remarkable advances that technology and science have achieved during the past 100 years and will continue to acquire at an exponential rate. Futurist, and a director of engineering at Google, Ray Kurzweil writes:

> *"By understanding the information processes underlying life, we are starting to learn to reprogram our biology to achieve the virtual elimination of disease, dramatic expansion of human potential, and radical life extension."*[7]

During the past century our life expectancy doubled. This increase can be attributed to better medical advances such as penicillin, a wide range of other improved medicines, vaccines, better medical procedures to treat illnesses, and improved living conditions.

If our current average life span (currently at 97 years) doubled again during this century, we would see many people living to close to 200 years. If this occurred, would we still need the religious beliefs of our ancestors, or would we begin to develop new beliefs? And regardless of what anyone thinks about religion, it is likely too ingrained in our existence to ever vanish completely.

Can We Re-Program Our Self-Imposed Suicide?

The problem we now face in terms of extending our lives is that the medical approaches that were used in the 20th century to solve diseases such as polio, diphtheria, and cholera may not be applicable in addressing aging and the diseases such as cardiovascular disease, cancer, and Alzheimer's that are associated with aging.

Our bodies are programmed to die. Our bodies are naturally set to self-destruct. Our natural defenses such as our immune system automatically shut down as we get older. Any attempts to significantly prolong our lives will be counter to the normal functioning of our bodies that are programmed to a self-imposed suicide as we age. Somehow, if we want to prolong our lives, we will need to find a way to turn off the self-destructive genes in our bodies and replace this with the programming that once kept us youthful.

While most of nature around us also experiences self-destruction, some plants and animals have managed to avoid aging. There are trees that are thousands of years old. Tortoises often live over 200 years. Clams are known to go beyond this with the longest living clam on record having survived for 507 years. Some animals, sharks as an example, while they have a fixed lifespan, appear to never age. And the granddaddy of longevity: yeast trapped in amber for 25 to 40 million years has been revived.[8] Is it possible that humankind with modern technological advances can match the longevity records of some other life forms?

Earlier in this book we looked at how natural selection impacts both pain and suffering. In the case of aging, natural selection has resulted in our bodies shutting down in a manner that is currently beyond our control. Our bodies are programmed to die. A question we could obviously ask is why. What benefits could aging hold for any life form, including us, on our planet? How did natural selection permit aging to occur when at first glance, aging would seem to be a contradiction to the concept of the survival of fittest?

How Does Death Benefit Our Species?

One of the more accepted current theories that answers this question focuses on the communal benefits that any species enjoys

WILL WE CREATE ETERNAL LIFE? / 135

as a result of the deaths of its members. For example, without death the community of any species would grow so large that overpopulation would become a rampant problem. It is possible that aging, while being a problem for individuals, might be very important to the overall health of any species. As we will explore in more detail later in this book (Chapter 10), natural selection may support cooperation within a species as well as sometimes being ruthlessly selfish for individuals.

Scientists, at the New England Complex Systems Institute and the Wyss Institute for Biologically Inspired Engineering at Harvard University, found that:

> "...*programmed deaths strongly result in long-term benefit to an organismal lineage by reducing local environmental resource depletion over many generations.*"[9]

Are There Problems Associated With Humans Living Longer?

With the advances in aging that we as humans have already achieved, there are both environmental and economic concerns associated with this. For example, who is going to pay for the rapidly increasing number of older people who are retired? There was a time when most people died within a few years after retiring. No longer is this the case. People in many industrialized nations now live 20 or 30 years beyond retiring. This places increased pressure on younger people who will directly or indirectly be contributing to the ongoing retirement benefits for older people. If people begin to live 40 or 50 years, or even much more, beyond retirement this could cripple the economy of countries where this happens.

In addition, as people live longer the world will become more populated. If the birth rate of humans (or any other species of animals) significantly exceeds the death rate, we will face a host of new problems as a result of overcrowding. As overcrowding occurs, there will be increased competition for food and other basic essentials. Such competition could become destructive, possibly even threatening the survival of our species. It is quite possible that throughout the billions of years of life on our planet, that natural selection favored the survival of animals who had a fixed limit on their lifespan because this contributed to the overall communal health and survival of the species.

An example of how overpopulation can lead to problems in a species would be the locusts (more specifically the Rocky Mountain Locust) that descended on mass in the mid-Western U.S. in the 19th century. Locusts are insects that resemble grasshoppers. In 1875, trillions of locusts, described as massive black clouds descending on the earth, invaded mid-Western U.S. states. As the locusts arrived, within hours they stripped the land of all vegetation including crops. It's estimated that the plague of locusts extended over an area that was greater than 2 million square miles. Some people described the locusts as a blinding black snow storm that covered the ground two to three inches deep and attached themselves to every inch of any creature, including humans, who might have been outside when the locusts came.[10]

Then unexpectedly, as suddenly as the horrendous mass of locusts had appeared, they vanished. The last recorded sighting of the Rocky Mountain Locust was in 1902. What happened to the locusts? Was the disappearance of the locusts an example of how overcrowding can lead to the demise of a species? Is there a lesson in this for humans?

We also need to be aware that while science may help us to live longer and likely have greater freedom from physical pain, the emotional pain of broken relationships, financial concerns, loneliness and failed dreams will remain part of our lives whether we live to be 60 or whether we live to be 260. As we have more success in postponing death, we will most certainly have to provide greater psychological care to address emotional pain that can accumulate with age.

Whether we die because natural selection found that it was in the best interest of our species to do so, or whether there is some other reason yet to be discovered, death has been programmed into our genes, and to overcome this reality is going to require some new approaches beyond exercise and healthy eating, although the current research most definitely shows that exercise and diet can positively impact our longevity (but nowhere close to what science may accomplish in the 21st century).

WILL WE CREATE ETERNAL LIFE? / 137

Evolving Technology

In the 21st century increasing scientific and medical discoveries are already beginning to address the problem of aging that is programmed in our genes.

Consider that personal computers and the Internet have been with us for less than 50 years. Five-hundred years ago it took a 100 years for the printing press to have a worldwide impact. Today, new advances (such as cell phones) can proliferate the planet within a decade, and in some cases new advances (think Facebook or Twitter) can spread like a hurricane in far less time). Who would have ever believed that we might someday have self-driving cars?

A 1949 article in Popular Mechanics predicted that in the future computers would advance to weighing only 1.5 tons.[11] Consider the small size of computers today and consider that most cell phones today are more powerful than a computer that would have filled a large room in the 1950s. The computing power of most cell phones today is greater than the computer power required to send the Apollo astronauts to the moon and back in 1969.

Intel executive David House (as a spin-off of Moore's law) predicted that chip performance would double every 18 months.[12] As computers continue to increase their performance (more on this in the next chapter—you might not believe where the future of computers is heading), their power to analyze complex problems will provide solutions that our human brains are incapable of solving, and some of the rapidly coming advances will relate to preventing aging and age-related diseases.

When the Human Genome Project started in the early 1990s, some critics predicted it would take thousands of years to complete. Thirteen years later the project was completed, and the exploration of our genome (as you will soon see) is progressing at a remarkable rate.[13]

How Soon Might We Double Our Life-Expectancy?

With increasing technological advances that will positively impact our health, is it possible that by the end of the 21st century we will have doubled our current life expectancy? Of course, there's always the possibility that climate change, pollution, overpopulation, a super virus (think plague), or a nuclear war could dramatically change these possibilities, but let's be optimist and consider the best of what might happen in the 21st century.

Is Eternal Life Within Our Grasp?

During the 21st century, there's a strong possibility that geneticists will discover which genes cause aging (there is already some success in this area in experiments with animals), and with such a finding will come the possibility of altering these genes to prevent getting older.

Is it possible that in this century medical advances will eradicate the major diseases of today (such as cancer), and we will have developed an inventory of body parts that might not be much different than going to an auto repair shop for a new heart or kidney? Is eternal life within our grasp, rather than being a fictitious hope that has permeated our past history? Is the elusive fountain of youth at hand?

In addition to genetic advances to prolong living, is it possible that technology will advance to the point that the content of a human brain can be downloaded into a computer, offering the possibility of a virtual eternal life?

Let's start our journey by looking at some current realities related to increasing the length of our lives and then explore some hypothetic possibilities of where we might be headed in our attempt to live forever.

Why Do We Die?

First of all, it would be useful to return back to natural selection which was discussed at length earlier in this book. From a purely physical standpoint we exist to pass on our genes. Once we reach a point where we are no longer (at least collectively) contributing to the welfare of our society and once we have reached a point where our continued living might threaten overpopulation, our genes are programmed to lead us to our demise.

Aging sets in in earnest. While diet, exercise, medical advances and improved living conditions may all deter death (as evidenced by the doubling of our life expectancy over the past century), there may still be an upper limit to our life expectancy that has been set in our genetic makeup. In order to push beyond these limits, it will be necessary for us to understand how our bodies are programmed to die. Once we understand this process, it will then be necessary to find solutions to counteract the body's attempts to self-destruct.

It's not my intention to provide a detailed account of how aging impacts our bodies, but a basic understanding of what happens to all of us would be helpful in considering how we as a species might begin to push beyond what are considered to be our normal life expectancy limits.

Senescence, or biological aging, is the gradual dying of our bodies, originating in the basic deterioration of our cells. Research and the understanding of senescence is progressing so rapidly that parts of this chapter could very well be out-of-date by the time this book is published, but none-the-less the information presented here will provide a glimpse at how our bodies age and how technology is beginning to look at how to overcome aging.

Everyone is prey to a silent internal assassin, a ruthless killer that dwells within which automatically causes us to self-destruct. As we get older, this killer begins to shut down our immune system as well as causing the deterioration of our cells.

Our genes control all cellular processes, including the division of our cells. It is normal for all of us to experience the death of millions of our cells every minute of every day, but these cells (at least in our youth) are replaced with healthy new cells. It is estimated that 50-70 billion cells die each day in every adult. As we age, our genes begin to lose their efficiency in replacing dead cells. In addition, our immune system begins to lose its power to fight against invading diseases.

What Are Some of the Causes of Aging?

Let's begin by taking a look at some current findings related to the causes of aging and then proceed to see how new medical advances may be used to disrupt or slow down aging. After looking at traditional illness treatments such as drugs, we will expand our thoughts to look at gene editing, gene therapy, and cloning. We will also look at some future possibilities that at the present time might be considered to be more science fiction than reality, but keep in mind that if you attempted to describe the Internet or a personal computer, that we now take for granted, to people 100 years ago, then you would have been most undoubtedly laughed at.

Sometimes the science fiction of today becomes the reality of tomorrow. As author Ray Bradbury writes:

> *"Science fiction is any idea that occurs in the head and doesn't exist yet, but soon will, and will change everything for everybody, and nothing will ever be the same again."*[14]

You are likely familiar with the fact that harmful plaque can form around our teeth. Without preventative and corrective measures, that plaque (which is a biofilm or mass of bacteria that grows on surfaces within our mouths) can become a problem. Plaque can lead to gum disease and can also cause demineralization of our teeth. As people get older, the problem of plaque increases and it takes a more concerted effort to disrupt and remove the mass of bacteria through brushing our teeth, using devices such as water picks, flossing, and professional cleaning. Fortunately, plaque can be easily reached in our mouths which makes it practical to both prevent to some degree and eradicate it when it does occur.

Similar to the concept of plaque impacting our gums and teeth, plaque can also build up in our arteries causing the inside of our arteries to narrow. In addition, plaque can build up on the neurons in our brain (known as beta-amyloid plaques). The plaque in our arteries can lead to atherosclerosis which includes heart problems, strokes, periphery artery disease and kidney problems. The plaque in our brains (along with another abnormal structure named tangles) can lead to diseases such as Alzheimer's (which is the most common form of dementia, accounting for 60-80% of dementia cases). At the current time it is thoughts that beta-amyloid plaque and tangles (twisted fibers of a protein called tau) damage and kill nerve cells in the brain, leading to diseases such as Alzheimer's, although genetic factors also impact the development of this disease.[15]

It's estimated that all of us have the beginning of beta-amyloid plaques accumulating in our brains by middle age which increase as we get older. As our brains accumulate beta-amyloid plaque, it becomes only a matter of time before our brains can longer function appropriately. While some treatments may temporarily improve symptoms, at the current time the damage caused to the nerve endings as a result of aging is irreversible.[16] By the time we're in our 80s, amyloid deposits occur across multiple tissues in our bodies.

As various forms of plaque build within our bodies, we then face a host of diseases. While improved diet and exercise can help to some degree in preventing concerns such as hardening of the arteries, in the end (pun intended), even the most athletic person in the world with the best diet will still experience death.

In addition to the concern with various forms of plaque building in our bodies, our cells begin to accumulate damage as we age, and we begin to experience mutations in our cellular components, primarily in those known as mitochondria. If we are to increase our life span significantly it will become necessary to find ways to prevent cell destruction from occurring in the first place (more on this later). While exercise, diet and medication can all contribute to a longer life, they can't at the current time reverse cell destruction and mutations which are at the heart of aging, and age-related deaths.

Aging is Not a Disease, but We Are Still Dying from It

As we look at current research into aging, we potentially face a huge problem, namely that aging is not in itself recognized as a disease. This results in a lack of government support and private funding to research approaches to slow down aging. Fortunately, research into diseases that are often age related (such as atherosclerosis, Alzheimer's, and cancer), will help to provide some solutions for aging as well as addressing these other diseases. For example, cancer is primarily the result of cell mutations. Strategies to cure cancer might also hold promise in addressing the cell damage that everyone experiences due to aging.

In many ways it is unfortunate that our governments (and even the general public) don't place a greater emphasis on research related to aging because aging is among the greatest known risk factors for most human diseases. It is estimated that two-thirds of the people who die each day across the globe die from age-related causes.[17]

It's as though we have simply accepted that we are all going to grow old and die without seriously considering that maybe things could be different. While there is a concerted effort across our planet to find cures for cancer (which in many forms is directly related to aging), there isn't the same emphasis on finding ways to prevent or interrupt aging which if successfully done could assist in eradicating cancer.

Ethical Concerns in Our Research Related to Aging

In considering research related to aging, studies often involve animals such as mice, fruit flies and even mold (and sometimes monkeys). While there are ethical considerations that surround such research, if we are ever going to find a cure for a disease such as cancer, clinical studies will begin with other animals before they progress to humans. This is generally an accepted part of the scientific process to find cures for diseases.

As you will see later in this chapter, the ethical considerations become even greater when we look at gene manipulation, cloning and stem cell research. Some countries have already banned or limited research in some of these areas based on ethical and religious concerns. Regardless of the moral stance that some countries may take related to advanced research in this area, other countries will permit and encourage such research. You might ask yourself if one country in the world successfully finds a way to significantly prolong life and eradicate diseases such as cancer, would you be willing to accept these discoveries even if they involve genetic manipulation, cloning or something to do with stem cells? Perhaps some of the information later in this chapter will help you to make a more informed decision

Immortal Life Forms

Interestingly we can find examples of other life forms that are immortal. For example, bacteria fission produces daughter cells. Strawberry plants clone themselves through runners. And animals in the genus Hydra can regenerate to avoid dying of old age.[18] Even within humans and some other animals, there are cells that appear to resist dying, such as cancer cells. Cancer cells have been maintained in cell cultures such as HeLa cell line.[19] In the artificial cloning of animals, adult cells can be rejuvenated to embryonic status and then used to grow a new tissue or animal without aging.[20] We will look at various forms of cloning later in this chapter as a possible approach to prolonging life.

WILL WE CREATE ETERNAL LIFE? / 143

Drugs as a Solution to Aging

For much of the 20th century and continuing into this century, drugs have played an important role in prolonging the lives of humans. The use of drugs in fighting age-related diseases and even aging itself will continue with even greater emphasis in the 21st century.

At the present time Elan Pharmaceuticals is working on a vaccine that would reduce the beta-amyloid plaque in our brains. Although their initial studies on mice were promising, after the U.S. Food and Drug Administration (FDA) granted permission to move the vaccine into placebo-controlled studies on humans, about one in fifteen patients developed a life-threatening swelling of the brain as a side effect.[21] Through further research, perhaps these dangerous side effects can be overcome. And keep in mind, if you've ever watched advertisements for accepted pharmaceutical drugs, that the list of side effects is often very lengthy for many drugs that are routinely used in our society. At some point the question becomes is the disease that the drug is eliminating a greater problem than the possible side effects a person might experience from taking the drug?

Experiments funded by the National Institute of Aging have shown that drugs can extend a mouse's life span by about 25%. If these same drugs worked on humans, in one quick move we would see our life span increase from the current projected average of 97 years to 121 years.[22]

Google Gets on Board with Longevity Studies

Tech giant Google has joined forces with researchers (with the launch of the company Calico) to look at longevity. Bill Maris, the person who is presiding over the Google Ventures investment fund for supporting Calico, in a January, 2013, interview said:

"If you ask me today, is it possible to live to be 500, the answer is yes."[23]

If scientists could discover a drug that prevents plaque from building in our arteries or in our brains, or a drug that prevents our cells from destructing or mutating as we get older, lifespan could be significantly increased. We have now entered a new stage in the history of humans when scientists are attempting to do just that.

One Pill Makes You Live Longer...

A drug called everolimus, used to treat certain types of cancers, was found to possibly reverse the immune deterioration that occurs with age. Novartis, one of the world's largest pharmaceutical companies, is currently exploring whether everolimus might reinvigorate the immune system in older people. At a 2014 conference, the head of Novartis stated that research into geroprotectors or longevity drugs was a priority.[24]

The drug rapamycin (also known as sirolimus) has been found to extend the lifespan of animals such as mice.[25] Various clinical studies have shown that rapamycin seems to increase lifespan as well as healthspan in mice. It has also been shown to have protective effects against many degenerative diseases caused by diseases such as Alzheimer's and cancer.[26]

Another drug, metformin (used for type 2 diabetes) has shown some promise in potentially prolonging lives. In a UK study of more than 180,000 patients, it was found that people on metformin were 15% less likely to die than people without diabetes who did not take the drug. A 2017 review and meta-analysis found that people with diabetes who were taking metformin had significantly less mortality from all causes, including cancer and cardiovascular diseases than those of other therapies.[27]

The FDA has given the go ahead for a trial to further study the possible anti-aging effects of metformin on humans. As Scottish aging expert Professor Gordon Lithgow of the Buck Institute for Research on Aging in California (who is one this study's advisors) said:

> *"If you target an aging process and you slow down aging then you slow down all the diseases and pathology as well."*[28]

Lithgow has also stated that the idea that we would be talking about a clinical trial related to aging in humans would have been inconceivable 25 years ago.

WILL WE CREATE ETERNAL LIFE? / 145

Researchers from the Mayo Clinic in Rochester, in 2017, asked for permission to test the use of senolytic drugs on humans, making the leap from animal testing. It is believed that senolytic drugs can kill problem-causing senescent cells (one of the major contributes to aging) without harming normal healthy cells. At the Mayo Clinic, researchers found that senolytic drugs can increase the length of time that mice stay healthy as they grow old. Dr. James Kirkland, director of the Kogod Center on Ageing at the Mayo Clinic, states:

> *"The same processes that cause aging seem to be the root causes of age-related diseases. Why not target the root of all these things? That would have been a pipe dream until a few years back."*[29]

The drug penicillin was discovered, largely by accident, by Scottish scientist Alexander Fleming in 1928. Penicillin and the related group of antibiotics have singlehandedly dramatically increased the lifespan of humans by fighting bacterial infections which up until the discovery of penicillin often led to death. Since its discovery, penicillin has saved many millions of lives. This one drug has made a huge difference, both in the elimination of pain and in the lengthening of human lives. What might be the wonder drug of the 21st century that dramatically impacts the lifespan of humans in the same manner as penicillin?

Kick-starting Our Immune System

As we age, our immune system weakens. Our immune system becomes less and less effective in combating infections and disease in our bodies. For example, influenza and pneumonia are rare in young people whose immune systems are healthy, but these diseases become a major cause of death for older people. Worldwide, lower respiratory infections (including pneumonia) are the 3rd leading cause of death in the world.[30] In the U.S. it is reported that 90% of all deaths for people over 65 years of age are the result of influenza and pneumonia.[31] Research also tells us that influenza in the elderly can cause heart attacks, strokes, and other disorders that appear to be unrelated to the respiratory problems.

Our immune system is a defense structure in our body that

fights against invaders. Back in Chapter 3 we looked at how there is a constant war in our bodies every minute as our bodies fight against infections. Unfortunately, as we age, our bodies begin to lose that fight. Our immune system begins to weaken. As well as the increasing dangers of influenza and pneumonia, the weakening of our immune system can also lead to autoimmune diseases, inflammatory diseases, and even cancer.[32]

Research in the 21st century will increasingly look at ways in which we can maintain a strong immune system even as we age.

Genetic Engineering

In addition to the search for a drug or drugs that may prolong our lives, let's also consider advances in genetics. While some people have a knee-jerk reaction to anything related to modifying our genes, research in this area is already occurring and genetic modifications in both plants and animals are a common and growing reality.

Some of the food we eat is genetically modified. Genetic engineering saves time and helps to produce bigger, higher-quality crops. If such crops are more nutritious and less dependent on pesticides and other chemicals, isn't that a good thing? If genetically modified crops could be helpful in reducing or eliminating hunger around the world wouldn't that be positive?

In Chapter 3 we looked at the difference between natural selection and selective breeding. Genetic engineering takes us a step beyond normal selective breeding as it allows the transfer of genes between organisms. For example, scientists have been able to insert a gene (known as cry1Ac) into tomato plants (and some other crops such as soybean, corn, and canola) to block beetles from devouring the plants. Cry1Ac is a gene from a bacterium found in the ground (Bacillus thuringiensis) that encodes a protein that is poisonous to beetles (and some other insects).

Genetic modification (GM) is also occurring in animals although the number of actual foods currently on the market from animals is limited. One GM animal that is available for human consumption is salmon which was approved by the American FDA in November, 2015.[33] The purpose of genetically modifying salmon was to create a fish that produced market weight in half the time, in addition to be-

WILL WE CREATE ETERNAL LIFE? / 147

ing able to grow these salmon domestically in artificially created ponds. The benefits of the genetically modified salmon would help to increase the availability of the fresh supply of a fish that has suffered a major decline in the wild.

According to the World Health Organization, genetically modified organisms (including plants, animals or microorganisms) are those in which genetic material (DNA) have been altered in a way that does not occur naturally by mating and/or natural recombination. It allows selected genes to be transferred from one organism to another and also between nonrelated species.[34] These are known as GM foods. Genetic modification offers a short-cut to selective breeding. Considering that it is estimated (by the UN Food and Agriculture Organization) that world food production must increase by 50% over present levels by the year 2050,[35] it's likely that we will see a significant rise in GM foods to meet this demand.

Cell Suicide and Aging

From previous chapters, you might recall that millions of our cells die every minute and are replaced with new cells. In studying aging, it has been discovered that every time a new cell appears throughout our lives (divides might be a better choice of words to describe what happens), the chromosomes (the long strands of DNA found in every cell nucleus which carry the blueprints for keeping us healthy) become a little shorter.

Over time, the shortening of the chromosomes in our cells leads to problems that are currently thought to be associated with aging. As our cells continually die and divide into new cells each day, to protect the chromosomes from losing key pieces of information that they are carrying, a small DNA buffer exists at the end of every chromosome to protect it. This buffer is called a telomere.

With every cell division that we experience, over many years of time, the telomeres begin to shrink which then makes them less effective in protecting the integrity of our chromosomes. Eventually, for all of us, our telomeres will shorten to the point that when our cells divide, the DNA within our cells will be endangered. When this happens, the cell enters a state known as senescence where it can no longer renew itself or carry out the work it was intended to do.

When a cell senses that the telomere is losing its effectiveness, it begins to age. The cell then becomes inactive, stops dividing, and can even poison younger cells around it. In a very real sense, the cell commits suicide (known as apoptosis) and in the process assassinates other healthy cells around it. As well, the senescent (aging) cells send out signals that cause inflammation throughout the body which is one of the primary causes of death. As the telomeres lose effectiveness, the stem cells that create new white blood cells slow down, thus weakening our immune systems.

If scientists could discover a way to keep the telomeres from losing their effectiveness, we might have a significant approach to prolong our lives, and also live healthier as we get older. Dr. Josh Mitteldorf, a theoretical biologist, in his book *Cracking the Aging Code* writes:

> "... people who are most prominent in telomere research tend to believe that telomerase will prove to be the philosopher's stone, the fountain of youth, the elixir of Gilgamesh about which humanity has dreamed for thousands of years."[36]

Royal Jelly

While there are some companies that sell natural herbal remedies (such as green tea, astragalus, and milk thistle) which might turn on the telomerase gene to repair telomeres, there has been little research done to actually support such claims. One study that showed some promise found that lifestyle changes including moderate exercise, stress reduction techniques, and a plant-based diet, increased telomere length by approximately 10%.[37]

An interesting lesson we might consider from nature is the queen honey bee. A queen honey bee lives up to 40 times longer than any other bees in the hive, yet she is genetically identical to all the other bees. The only difference is that this bee is chosen to eat a special diet of large quantities of royal jelly produced by the bees. This royal jelly causes a reduction in DNA methylation which in turn activates the genes associated with royal jelly.

If we were able to increase the lifespan of humans 40 times like a queen bee, we would prolong our lives to over 3,500 years. Could dramatic increases in prolonging the lives of humans be as simple as

WILL WE CREATE ETERNAL LIFE? / **149**

someone discovering a royal jelly for humans? Royal jelly from bees is actually sold as a dietary supplement, but the European Food Safety Authority and United States Food and Drug Administration have concluded the current evidence does not support the claims of any health benefits.[38]

Gene Editing and Gene Therapy

During this century, it's quite possible that telomere repair might be accomplished through new drugs, gene therapy, or metabolic suppression. In order to make significant advances in repairing any genes that cause aging, it will likely occur in the area of gene editing or gene therapy.

Gene editing might be defined as repairing defective genes, while gene therapy attempts to fix genetic problems by adding healthy genes.

Telomere extension in laboratory mice at Harvard has successfully shown a reversal in some signs of aging.[39] These mice were genetically engineered with a chemical on/off switch to control telomerase which is thought to have the potential to transform telomeres to a more youthful state. In these experiments, mice with an extra copy of the telomerase gene inserted into them, lived longer.

There is already a buzz on the Internet about gene therapy being used to lengthen telomeres in humans. Offshore clinics, as well as possible unregulated clinics in some countries, may already be conducting gene therapy with telomerase.

We all have approximately 3.2 billion letters that make up our human genome, a genome being any organism's complete set of DNA, including all its genes. Each genome contains all the information required to build and maintain an organism. Sometimes what are called the individual letters within our genome have letters that are defective. These defections can cause diseases. They can also result in unusual physical characteristics, even impacting our intelligence and personality. As we age, more defections occur leading to greater health problems which then cause death.

Consider a book editor who scans a book for grammatical, spelling and contextual errors (not too many here, I hope). With skilled fingers on a keyboard an editor can quickly correct any mistakes. The technology now exists to allow a gene editor to target and correct mistakes in our genome. DNA is like a secret language with sequences of letters that provide instructions to build and maintain our bodies. We have now reached a new threshold in human history where we can edit the language that has created us.

Designer Babies

Perhaps you have heard of CRISPR (an acronym for "clustered regularly interspaced short palindromic repeats"). CRISPR, as one of the most successful developments in gene editing, could help scientists to manipulate and modify the genetic code of any creature on our planet. This technology has the capacity to correct or repair mutations within our genes which might cause various diseases (and this technology might eventually lead to altering the genes which control aging).

While technology such as CRISPR may help to solve major diseases such as cancer or even correct many hereditary diseases, this might also lead to attempts to create "designer babies" with perfect looks and high intelligence. Some fear that gene editing could result in an attempt to create a super race of humans. While once the arena of science fiction, this is now becoming closer to a very real possibility in the future (more on this later).

In 2017, researchers in both China and the U.S. succeeded in removing a disease-causing gene (an inherited heart condition) from human embryos.[40] If these studies continue (which they most certainly will even if the U.S. shuts down or severely limits such research), the next step would be to place the gene-edited embryo into a woman, potentially giving us a genetically engineered child. As further advances occur in gene editing, it's quite possible that the genes that trigger aging, once identified and fully understood, may also be edited to significantly prolong life along with improved health in old age.

CRISPR

Considering that a technique like CRISPR (the term was first used in 2002) is basically a 21st century invention, the advances that have already been made suggest a future of staggering possibilities. Research into our genome is progressing so rapidly that you can now, fairly inexpensively, order a personal DNA sequence analysis

WILL WE CREATE ETERNAL LIFE? / 151

simply by providing a saliva sample from inside your mouth.

It should also be noted that CRISPR is not the only gene editing strategy. In addition to CRISPR there is also ZFNs and TASLEN.[41] It is very likely in our rapidly advancing technological world that new advances in gene editing will continue to appear.

As individuals begin to explore their own DNA for possible genetic errors, how long will it be before human embryos are routinely tested while still in the womb, and parents are provided with choices to repair faulty genes before their child is born?

CRISPR-Cas9 is a family of repeating DNA sequences in bacteria cells. These snippets of DNA can be used to modify genes so that genes which may be causing medical problems can be removed and new genes added. CRISPR has already been used to modify the DNA of pigs in an attempt to create pigs whose organs might someday be used as donor parts for humans. CRISPR is also being used to manipulate the genes of mosquitos to possibly eradicate diseases that some mosquitoes carry such as malaria and Zika.

Have We Replaced God?

In the book *A Crack In Creation* by Jennifer A. Doudna, a professor in the chemistry and the molecular and cell biology department at the University of California, and Samuel H. Sternberg, a biochemist, we read:

> *"It amazes me to realize that we are on the cusp of a new era in the history of life on earth—an age in which humans exercise an unprecedented level of control over the genetic composition of the species that co-inhabit our planet. It won't be long before CRISPR allows us to bend nature to our will in the way that humans have dreamed about since pre-history. When that will is directed toward something constructive, the results could be fantastic—but they might also have unintentional or even calamitous consequences."*[42]

Jan van Deursen, a biochemist at the Mayo Clinic in Minneapolis, has inserted a gene into mouse embryos which had a receptor in it for a drug known only as AP20187. When cells began to age in these mice, before they could cause the destruction of other cells, the AP20187 killed the aging cell but didn't touch the healthy cells. The results produced mice that were healthier. Their kidneys were healthier, and their hearts were more resilient to stress. As a result, the genetically engineered mice lived 20-30% longer than normal.[43]

Researchers at the Salk Institute for Biological Studies modified genes by turning adult cells back into embryonic-like ones, in mouse and human cells in vitro. These results led to the conclusion that it is possible to slow down or even reverse aging, at least in mice.[44]

At the Harvard Medical School, Dr. George Church, has already been able to reverse aging in mice and in human cells using CRISPR. Dr. Church's company, Editas Medicine, is already being backed by Google Ventures and the Bill and Melinda Gates Foundation. Dr. Gregory M. Fahy, after hearing a presentation by Dr. George, stated the following:

> *"If aging is controlled by master genes, and if the activity of such genes can now be intentionally controlled, then we are beginning to approach the control of aging at a very fundamental level. And the same technology can be applied to the correction of many diseases as well, whether age-related, or not."*[45]

A New Way to Fight Cancer

In addition to gene editing, gene therapy may also play a role in extending our lives. Gene therapy involves inserting a gene into a patient's cells instead of using drugs or surgery. This process might result in replacing a mutated gene that is causing a disease with a healthy gene, or it might involve destroying a mutated gene that is causing an illness.

Immunotherapy, a form of gene therapy, attempts to supercharge the body's natural defenses in fighting disease. In the U.S., as an example, a living drug named tisagenlecieucel is the first gene therapy of any kind to be approved by the FDA. Tisagenlecieucel will be used to treat certain forms of leukemia in children and young adults who have stopped responding to chemotherapy. In this form of therapy, white blood cells (known as T cells) are collected from

WILL WE CREATE ETERNAL LIFE? / 153

the patient's blood. Next, genes that recognize specific cancer cells are inserted into the T cells using an inactive virus. After the T cells are grown, they are infused back into the patient where they multiply and begin to hunt and kill the target cancer cells. Researchers are now testing similar forms of gene therapy against other forms of cancer.[46]

If we are to successfully prolong life in a significant manner, cancer is one disease that we will have to overcome. Some researchers have stated that if humans lived long enough, we would all end up getting cancer.[47] If gene therapy and gene editing could eradicate this crushing disease, it would be a great victory for humankind. Gene therapy and gene editing might very well contribute to solving the problem of cancer as well as some other diseases.

In addition to looking at a cure for cancer, it is estimated that hundreds of clinical trials are currently occurring in the U.S. using gene therapy to defeat diseases such as hemophilia and cystic fibrosis.[48]

Replacing the Genes that Cause Aging with Genes that Emphasize Youth

It's possible that gene therapy could be used to kill mutated genes in our body that are the result of aging or that cause age-related diseases. It's even possible that the genes that cause aging could be replaced with more youthful genes. Such a possibility could significantly prolong our lives.

Dolly (not Parton)

In thinking about the creation of eternal life as a result of scientific advances, let's move on to cloning. In a newspaper this morning I read the following article headline, "Scientists use the Dolly method to clone monkeys."[49]

Perhaps you might be aware that Dolly, a sheep, was the first mammal (but not the first animal) cloned back in 1996. In cloning Dolly, scientists took a cell from a mammary gland, implanted this cell into an unfertilized oocyte (developing egg cell) that had its cell nucleus removed. The cell was stimulated to divide by using an electric shock (sounding a bit like the creation of Frankenstein's monster?). When it developed into a blastocyst (a structure formed in the early development of mammals), it was implanted into the surrogate mother. Since Dolly's time, pigs, dogs, deer, mice, frogs and horses have all been cloned. Although Dolly's legacy may have contributed more to stem cell research than cloning, we will save that topic for a little later in this chapter.

As the article I was reading this morning stated, creating two healthy monkeys (by cloning) brings science an important step closer to being able to do the same with humans.

Perpetual Life in Nature

While the thought of cloning brings trembling horror to the minds of some people, nature is full of examples of cloning. If you have ever grown strawberries you have likely seen runners branching out from the parent plant which form new cloned plants. The strawberry plant is a perpetual life form that if it were not for disease or wanton destruction (generally by humans) what began as a single plant might keep cloning itself forever.

Cloning is the process of producing genetically identical individuals or organisms as a result (in nature) from asexual reproduction. Bacteria, some plants, some insects, and even some animals can clone themselves. If you've ever taken a cutting, such as a branch from a plant, and immersed it in water for a period of time while you wait for roots to develop, you are cloning this plant.

Sea sponges naturally clone themselves through a process known as gemmulation. Tapeworms, upon reaching maturity, detach part of their body (known as a proglottid) to clone themselves. Aphids, those pesky little bugs, that can destroy plants, self-replicate themselves. Whiptail lizards can replicate themselves through cloning. There are even verified reports of both sharks and snakes giving birth to young without ever having a sexual partner.

The marbled crayfish is one of the most amazing species of animals in our world. It is reported that 25 years ago, this species of crayfish did not exist. Then a genetic mutation not only created this new species, but it also allowed the female crayfish to clone itself which has resulted in millions of these crayfish appearing in European lakes and rivers.

Male Sperm Counts Dropping

With declining sperm counts in male humans across the planet (halved in the last 40 years), will sexual reproduction among humans evolve into something different than what we experience now? Will scientists discover something from the marbled crayfish to help humans reproduce in a totally new way? An article in the BBC News Health Section stated, Dr. Levine, an epidemiologist, was quoted as saying:

> *"If we do not change the ways that we are living and the environment and the chemicals we are exposed to, I am very worried about what will happen in the future. Eventually we may have a problem, and with reproduction in general, and it may be the extinction of the human species."*[50]

Born Again

Human cloning might result from taking a cell sample from somewhere inside the body of a human and placing it in an unfertilized egg with its nucleus removed and then inserting this egg (after it has been coaxed into forming a blastocyst) into a surrogate mother (if we were to use the approach that was used to clone Dolly).

Using this method, an aging person could have a literal "twin" created. Repeating this process could result in a person living theoretically forever. The problem, of course is that each successful twin would not have the mental memories of the person they were cloned from.

Perhaps the day will come when humans can be recreated through cloning and have the data from their former brains downloaded into their new brain (of course, this would result in a host of problems if the content of an adult brain was downloaded into a cloned twin who was now a baby).

In the June, 2015, issue of *Medical Daily*, Dana Dovey writes:

> *"While the ethics and legality of human cloning are blurry, the science behind the idea is quite clear, with all research suggesting the practice is quite possible. Scientists have already cloned human*

embryos and many believe that creating fully developed humans is the next step."[51]

Stem Cell Controversy

As scientists have conducted research into cloning, this has opened another area that could have huge implications related to prolonging our lives, an area that is so controversial that some governments have placed restrictions on this research. This is stem cell research.

In 2000, the Clinton administration issued guidelines that allowed the federal funding of embryonic stem cell research.[52] In 2001, the George W. Bush administration restricted federal funding or research on stem cells obtained from human embryos. In response to this legislation President Bush said:

> *"At its core, this issue forces us to confront fundamental questions about the beginnings of life and the ends of science. My position on these issues is shaped by deeply held beliefs. I believe that human life is a sacred gift from our creator."*[53]

Less than two months after taking office after Bush, President Obama lifted the restrictions on federal funding for stem cell research.

What Are Stem Cells?

What exactly are stem cells and how might stem cell research aid in prolonging our lives?

The cells in our bodies each have a specific purpose. As a fetus develops in the womb, specific cells contribute to the growth of specific body parts. For example, some cells grow into arms while others grow into our kidneys or even our heart. As adults our bodies still have stem cells that can help to repair or grow new tissue.

In the early stages of a pregnancy, after the sperm fertilizes the egg, an embryo forms. Three to five days later, the embryo becomes a ball of cells known as a blastocyst. A blastocyst is made up of an inner cell mass and an outer cell mass. Stem cells, that are found in the inner mass, have the potential to become any body part for about five days before the embryo implants itself into the uterus. Once the embryo enters the uterus or womb, the stem cells begin to differentiate which means they begin to form specific body parts. These stem cells are known as embryonic stem cells.

While most people tend to think of embryonic stem cells when they hear anything about stem cells, stem cells known as mesenchymal stem cells (MSCs) can come from many tissues including fat and bone marrow in adults, and umbilical cords left over from child birth. Using adult stem cells (and the success rate is higher here when a person is using his own adult cells) or those from donated umbilical cords after a birth (or the amniotic fluid that surrounds the baby in the womb, or even the placenta after birth) eliminates the concerns that have arisen related to taking stem cells from the embryos of unborn children.

At the current time, embryonic stem cells appear to have much greater potential in treating health problems, but the problem is that taking stem cells from a human blastocyst prevents the fertilized egg from forming into a human. As a result, researchers are attempting to find other ways to retrieve stem cells which includes taking them harmlessly from the umbilical cord blood after delivery. In addition, scientists could take stem cells from fertilized eggs in in vitro fertilization (IVF) where doctors fertilize several eggs to ensure that one survives. As a result of this procedure some fertilized eggs are not used so they are either frozen or destroyed. If they are going to be destroyed, they might be donated to stem cell research, but this is a controversial area.

Let's consider one example of stem cell research to illustrate its potential. In 2016, there were 121,678 people in the U.S. alone waiting for a kidney transplant. In 2014, 4,763 patients died while waiting for a kidney transplant.[54] If scientists could learn how to grow kidneys (or other body parts) from stem cells, a significant number of lives could be saved.

Religion and Stem Cells

Although there is often an outcry from conservative religions regarding stem cell research, the Catholic Church supports adult stem cell research (which includes the use of umbilical cords donated after the birth of infants). In 2006 Pope Benedict XVI, after a symposium organized by the Vatican, gave support for adult stem cell-based therapies. In 2005, the Catholic Church gave more than 10 million dollars for stem cell research,[55] and in 2013 it gave further financial support in partnership with a U.S. based stem cell company (NeoStem) to fund research and education on adult stem cells.[56]

Stem Cells and the Fight Against Cancer

In cancer patients, the use of chemotherapy designed to destroy the cancer cells can also destroy bone marrow, reducing the patient's ability to make new blood cells. Donor blood cells to resolve this concern are often rejected because the body reacts because they are foreign. On the other hand, stem cells from umbilical cords are not perceived by the body as being foreign, therefore the immune system does not reject them. Further research in this area holds great promise in successful bone marrow transplants which could aid in fighting cancer.

Stem Cells and Aging

Stem cells are found naturally throughout our entire bodies. They live in a specific place in our tissues, often remaining dormant until a disease or injury activates them to repair tissue. Stem cells routinely help to repair damage such as cuts, broken bones, sprains, muscle tears, inflammation, and so on. Unfortunately, as we grow older, our stem cells diminish in number. Eventually we don't have enough stem cells to prevent age-related diseases.

We have two main types of stem cells in our body. We have MSC's (which have already been mentioned earlier in this chapter) that are found in every tissue. In addition, we have EPC's (endothelial precursor cells) which are primarily found in our bone marrow and circulate in our bloodstream.

Replenishing Our Stem Cells to Slow Down Aging

A simple way that most people can increase the circulation of EPC's is through cardiovascular exercise and a diet that is high in antioxidants (such as blueberries or green tea) which in turn can help us to live healthier and slightly longer.

Caloric restriction is also an easy and practical way to increase longevity and health. Studies, since the 1930s, have routinely showed the caloric restriction can be effective in fighting aging. While exercise and diets can increase the flow of stem cells in our blood, they don't actually replenish our supply of stem cells. Although we may live slightly longer as a result of diet and exercise (some studies suggest up to 20% longer) and we might live healthier, we will still not achieve eternal life.

A more significant approach to prolonging aging might be the use of cultured stem cells from umbilical cords to replenish our stem cells as we get older.

Cultured umbilical stem cells are currently being researched for use in a wide variety of diseases. They have been used in the effective treatment of leukemia[57]. It is also reported that cultured umbilical stem cells may be effective in improving some functions in some children with cerebral palsy. Other studies have reported the effective use of umbilical stem cells in repairing tissues in other animals after heart attacks and strokes. As stem cell research is ongoing it may be some time before we know the full benefit of any applications in humans.

MSC's have shown to be effective in prolonging the lives of rats in clinical studies. A group of Korean researchers used human MSC's in rats. As a result of the MSC treatments, the life span of rats increased 23-31%.[58]

In looking at aging, mesenchymal stem cells (MSCs) which exist naturally throughout our bodies begin to decrease in effectiveness over time, making it more difficult for our bodies to fight infections and diseases such as cancer. A question we might ask here is it possible that a MSC supplement (in the form of a pill or even a vaccine) could be taken as we age to boast our natural systems for fighting age-related diseases?

Perhaps the first step was taken in this direction when the world's first approved stem cell drug was brought to market in Canada and New Zealand (and later approved by the FDA in the U.S. in 2017) for the treatment of GvHD in 2007 (GvHS is known as graft vs host disease which is basically a harmful reaction after bone marrow transplants that kills around 80% of the children affected).[59]

Another possible use of MSC in slowing our aging process relates to the health of our cells. Cell robustness contributes to living longer and living healthier. Every day, 50-70 billion cells die in an average person and are replaced with new cells. Every second, 2.5 million red blood cells die in each of us. As we age, our body loses its ability to constantly replace the dead cells. It's possible that the therapeutic use of MSC could help to increase our cell creation as we age.

Dr. Robet Hariri, Co-Founder and President of Human Longevity Cellular Therapeutics and Founder, Chief Scientific Officer of Celgene Cellular Therapeutics, in an interview, stated the following:

> *"What if we could capture stem cells with our original, uncorrupted DNA at birth and then replicate them into a large number of future dosages and then freeze those dosages, making them available for future injection to facilitate repair over the course of our lives."*[60]

As of April 2016, there were more than 500 MSC-related clinical studies on the NIH Clinical Trial Database. The various forms of research studies related to MSC were looking at a wide range of therapeutic applications. These applications included the use of MSC in treating injuries and/or inflammation with 42% of all clinical trials related to immune-/inflammation-mediated diseases (which could prove to have a direct link to the problem of aging). Some of these research studies also looked at the use of MSC in organ transplants. Twenty-one per cent of these trials used MSC obtained from fetal sources such as umbilical cords, umbilical cord blood, and placenta. Some of these trials looked specifically at using stem cells for the treatment of graft-versus-host disease, multiple sclerosis, osteoarthritis, rheumatoid arthritis, inflammatory bowel diseases, and pulmonary diseases.[61]

Growing Human Organs in Sheep and Pigs

Recently, human-sheep hybrids have been created in the laboratory which could possibly lead to growing an unlimited supply of organs for humans awaiting transplants. In this ongoing research, human and sheep cells were grown inside a surrogate for three weeks, resulting in embryos that had both human and sheep cells.

WILL WE CREATE ETERNAL LIFE? / **161**

The next phase of this research would be to implant human cells into sheep embryos with the eventual goal of growing human organs in animals such as sheep. This same process is also being done with pigs.

Dr. Hiro Nakuachi, one of the leaders of the research involving the human-sheep research has stated:

"Organs grown in animals will be available for transplant within the next five to ten years."[62]

Could Genetic Engineering Create New Weapons of Mass Destruction?

In considering the remarkable possibilities that gene editing, cloning and stem cells might make in fighting diseases and prolonging life, we should also pause to consider the flip side of these potential life enhancing possibilities.

In 2016, James Clapper, the U.S. Director of National Intelligence, named genome editing as a potential weapon of mass destruction.[63] In the same way that gene editing could be used to eradicate disease carrying mosquitos, the same techniques could be used to target and eliminate specific races of humans. We might very well ask how a ruthless dictator like Hitler might have used any of these technologies to cleanse the world of those who lacked the characteristics of his master race and how he might have used these advances to create his genetically pure master race.

In Aldous Huxley's Brave New World (set in 2540) the author depicts genetic castes. Without careful monitoring and guidelines, we might find ourselves entering Huxley's world long before 2540.

Do We Have An Internal Clock that Controls when We Age?

Have you ever woken up before your alarm clock startled you? Most of us have experienced this. I tend to wake up about 30 minutes before I have to get out of bed. I enjoy this quiet time when I'm partially asleep and partially awake.

During this time I'm often aware of our cat attacking phantom menaces around our house, but an interesting thing happens a minute or two before I actually get out of bed—our cat comes and sits beside my side of the bed, knowing that it's about to get fed (and yes, you might be quite right if you're thinking that perhaps our cat is really my alarm clock).

For those readers who have dogs or cats, you're likely well aware that these pets often know exactly when family members usually wake up and they often anticipate a family member coming home a few minutes before they arrive (if there is a specific routine to the arrival of the person, and especially if it's a person like my wife who lovingly pampers our cat the moment she steps inside the house).

It appears that we all have internal clocks that contribute to us waking up in the morning (although some people have found a subconscious way to shut this clock off). This biological clock can also have us yawning when it's time to go to bed. Such a clock appears to be present throughout nature.

There are flowers that open when the sun rises and close when the sun sets. There are animals that sleep when the sun rises and wake up when the sun sets. Nature is also full of examples of an internal clock that impacts behavior as the seasons change. In some countries the leaves fall off trees in the fall. In some countries, animals almost magically seem to know when to start storing food because winter is coming, and they are going to hibernate.

At least for humans, and likely for our household pets, a clump of nerves in our brain (called the suprachiasmatic nucleus), oversees our internal clock. This is known as the circadian rhythm. Our sleep cycles are regulated by PER which is a protein found in our body. Circadian rhythms have been found in plants, animals, and even fungi.[64]

A question we might ask related to circadian rhythms might be: "If we have an internal clock that controls our waking and sleeping patterns, is it also possible that we have another internal clock that sets the whole aging process in motion?"

Throughout this chapter we have explored the various ways that the body begins to age. It has been consistently noted that the body is programmed to age. The various scientific and medical strategies we have looked at to prolong our lives (and live healthier) have generally focused on attempting to interrupt or alter the aging process in our bodies. What if the aging process is controlled (likely in our genes) by a switch like a thermostat on the wall of our homes? Is it possible that in the same way that all of nature has an internal clock that regulates waking and sleeping, that we might have a similar clock that gives instructions to the body when it's time to begin to age (or time to enter puberty, or time to enter adulthood, or a time to reproduce)?

If we have an internal clock located somewhere in our genes that provides our bodies with a set of instructions related to the timing and speed of our aging, is it possible that the identification of such genes could hold the key to finding the fountain of youth?

How Can Nanotechnology Prolong Our Lives?

Finally, as we consider our search for eternal life (or at least the prolonging of our lives), let's consider nanotechnology. Most people tend to think of very tiny robots or miniature mechanical devices when they think of nanotechnology. You might be familiar with science fiction movies or books that consider the possibility of tiny nanobots entering a person's bloodstream to provide some form of surgery. For example, the movie *Fantastic Voyage*, released back in 1966 considered such a futuristic possibility as did the movie *Innerspace* and the TV show *The Magic School Bus*.

Nanotechnology is the manipulation of matter on a very small scale (atomic, molecular, supramolecular, and quantum-realm scales). Nanotechnology involves working with individual atoms. A nanometer is defined as one billionth of a meter (a strand of human hair being one/one-hundred thousandths of a meter).

Christina Peterson, the co-founder of Foresight Institute, states:

> *"Nature manipulates individual molecules to build most complex things in the world—plants, animals, and our own bodies. The goal of nanotechnology is to use systems of molecular machines to build whatever we want with that same level of precision, and do it cleanly—just as nature does."*[65]

Nanotechnology research is currently receiving billions of dollars from both private and government sources. While there are a wide range of military and industrial uses in the diverse field of nanotechnology, research is also being conducted in its application to our health and longevity. In 2005, as an example, the National Cancer Institute in the U.S. committed 144.3 million dollars to fund research on the application of nanotechnology in fighting cancer.[66]

At the current time, the application of nanotechnology is being explored in the fight of diseases such as cancer, diabetes, AIDS, cystic fibrosis, malaria, and heart disease. It is estimated that there are more than 4,000 diseases that have a genetic component, making all these potential targets for gene therapy involving nanotechnology.[67]

Robert A. Freitas, a senior research fellow at the Institute for Molecular Manufacturing, speaking at the Fifth Alcor Conference on Extreme Life Extension, said:

> *"The net effect of these nanomedical interventions will be the continuing arrest of all biological aging, along with the reduction of current biological age to whatever new biological age is deemed desirable by the patient, severing the link between calendar time and biological health. Such interventions may become commonplace several decades from today. Using annual checkups and cleanouts, and some occasional major repairs, your biological age could be restored once a year to the more or less constant physiological age that you select. You might eventually die of accidental causes, but you'll live ten times longer than you do now."*[68]

What is the current reality of nanotechnology related to fighting disease and aging? Could a tiny device be implanted in a person's blood stream to perform surgery? If so, could a similar device be implemented in a person's bloodstream to clean the plaque on his arteries to prevent heart disease, or to clean the plaque on the neurons in his brain to prevent Alzheimer's? The answers to these questions, and the future of nanotechnology, are more incredible than you might ever imagine.

How Nanotechnology Could Fight Cancer

Previously in this chapter we looked at how gene therapy and gene editing may provide answers to fighting diseases such as cancer or altering our bodies built-in death programming.

Nanotechnology can contribute to gene therapy by helping to manipulate genes on a cellular level. By doing this, a disease can be stopped before it spreads. Nanotechnology can help target drugs or gene therapy to fight against disease in specific locations in the body.

For example, chemotherapy is often used in the fight against cancer. One of the problems with chemotherapy is that this form of therapy attacks healthy cells in addition to the cancerous cells. Chemotherapy can cause undesirable side effects such as damage in the heart and bone marrow. Using nanotechnology, the hope is that eventually the diseased cells, caused by cancer, can be attacked directly without causing unwanted side effects. In such a situation, nanoparticles would deliver the drugs directly to the diseased cancer cells.

It's even possible that strands of DNA could be injected into a cancerous area with the use of nanotechnology, and these DNA strands could kill the cancer cells without the use of chemotherapy. Nanotechnology will be one more tool in helping to eradicate diseases at the genetic level without the possible side effects of past therapies.

Nanotechnology and the Early Detection of Diseases

Nanotechnology may also be involved in detecting diseases at an earlier stage. This would make it easier to stop a disease before it becomes full-blown. At Ohio State University, engineers have developed polymeric nanoparticles filled with even smaller particles (called "quantum dots") that can help scientists to observe the development of cancer at a molecular level, providing information that could help to treat the disease.[69]

Swapna Upadhyay, a research scientist from the Institute of Environmental Medicine in Stockholm, Sweden, states:

"Application of nanotechnology has significantly increased in different spheres of life including the drug delivery systems and is being considered to be the technology of the future."[70]

Nanotechnology Apps

In the same manner that cars now have a computerized diagnostic system, is it possible that we are not too far away from having nanobots travelling throughout our body to monitor our heath? Is it also possible that eventually these nanobots will fight any diseases they discover or at least help to target various therapies to the exact site of the disease? Will these nanobots be able to manufacture healthy cells or even eradicate genes which cause diseases or aging?

More and more people wear "smart" watches or bands to measure their heart rate, and the results of various forms of physical activity. There are various apps that can be used in conjunction with these devices.

We are quickly approaching the day when these smart bands or our cell phones may communicate with nanobots floating in our blood stream. As a result, we would have instant access to the identification of diseases developing in our bodies. As well, the future will likely bring nanobots that will act like miniature physicians in fighting diseases in our body.

Perhaps the day will come soon when a person who suffers a heart attack or a stroke will already have an implanted nanobot that begins to treat the emergency before other medical help arrives (and such a nanobot could also place an emergency 911 call for help).

In addition to contributing to our health and longevity on an individual basis, technology may hold the key to fighting worldwide environmental concerns. Nanobots could be used to clean up pollution, even potentially rebuilding our ozone layer. With nanotechnology comes the possibility of creating building materials that don't require us to diminish the natural resources on our planet.

Chapter Summary

In this chapter, we have looked at a myriad of technological and scientific developments in the 21st century that will prolong our lives and help us to live healthier. A basic premise in this chapter is that death was programmed into our genes because this was in the best interest of the survival of our species. Now, that our understanding of our genetic makeup is rapidly increasing, we have the potential to modify our genes which could result in the eradication of major diseases such as cancer. In addition, genetic engineering

WILL WE CREATE ETERNAL LIFE? / 167

could help us to slow down or even reverse the body's self-inflicted suicide as we get older.

We are very quickly developing the power to create life in whatever direction we want go. As humans become God-like, how will this change our search for meaning? In Chapter 11, we will explore this further. Let me end this chapter with a quote from Yuval Noah Harari in his best-selling book *Home Deus*:

> *"Just try to imagine Christianity, Islam or Hinduism in a world without death—which is also a world without heaven, hell or reincarnation. Modern science and modern culture have an entirely different take on life and death. They don't think of death as a metaphysical mystery, and they certainly don't view death as a source of life's meaning. Rather, for modern people death is a technical problem that we can and should solve."*[71]

- 8 -
Will Virtual Reality Allow Us To Create Our Own Narrative?

When I was in my twenties, two of my friends were involved in a serious car crash late one Saturday evening. I received a call about this terrible accident shortly after my two friends were admitted into an emergency ward at a local hospital. Both Jim and John (not their real names) suffered severe head injuries. After I arrived at the hospital, each of their parents told me that the prognosis was not very good.

Jim died before morning while John remained in a coma for another eight months before dying. I remember visiting John from time to time, watching his body begin to shrivel. Some people said that Jim was the lucky one.

John's parents read to him regularly. Friends visiting John told him what was happening in the world as though he was completely coherent, although he never reacted to our presence. It was painful to watch him never move. It hurt to realize he would likely never recover. It was emotionally draining to watch those who were closest to him suffer. Eight months after the accident when he died, although I felt grief from losing a friend, I also felt relief that the end had come.

I was asked to speak at the funerals of both of my friends. Standing in front of hundreds of young people filled with the anguish and distress that death came far too soon for two of our friends, the devastated parents of each boy, and the distraught siblings deeply challenged my core beliefs about the meaning of life.

Our Brains Are Who We Are

Adrian Owen, currently the Canada Excellence Research Chair in Cognitive Neuroscience and Imaging at the Brain and Mind Institute at Western University, has discovered that some people who are in a vegetative state after suffering a serious head injury (or even a stroke) are actually still mentally alive. Can you imagine the incredible disturbing range of emotions a person—in what was thought to be a vegetative state—might experience if they could understand everything that was happening around them, but could not communicate this to anyone else?

Owen's research has led him to communicate with patients who were thought to be brain dead. In a very real sense, some of these people were caught in their own dreams, and making it worse for some of them, the people around them (such as family, friends, and medical staff) often didn't realize that they could understand everything that was being said.

While some of the people that Owens researched recovered, others died before regaining visible consciousness. He describes one woman who recovered but said she had wanted to commit suicide when she was in a vegetative state unable to communicate with anyone around her. She said that she even tried to hold her breath, but her body prevented her from killing herself.

In his book *Into The Gray Zone*, Owens writes,

> *"The most important lesson I learned is that your brain is who you are. It's every plan you've ever made, every person you've fallen in love with, and every regret you've ever had. Your brain is all there is. It's the pulsating essence of you as a person. Without a brain, our sense of 'self' is reduced to nothing....I learned the most important lesson in neuroscience—we are our brains."*[1]

A Personal Tragedy

More than two decades ago, my precious fourteen-year-old son Tom was hit by a car while crossing the street on his bike. According to the eventual police investigation of the tragedy, the car that hit him was traveling at well over twice the speed limit in a residential neighborhood.

It was the second week of September. Tom had entered grade 9 at a nearby high school. Tom and many of his friends were elite hockey players. I was to be Tom's hockey coach for the upcoming season. Actually, I was sitting in our family room, completing applications for various hockey tournaments when there was a frantic pounding at my front door.

When I opened the door, one of Tom's friends exploded with, "Tom has been hit by a car."

These were words that no parent ever wants to hear. I stood in shock before reacting.

Within minutes, I rushed to the scene of the accident to find Tom lying in the middle of a street, the medics and firemen attempting to resuscitate him. I held Tom's hand. There was no life.

Once the emergency crew was successful in starting Tom's heart, he was rushed to the hospital. I held his hand in the emergency ward—there was still no life. The agony was so overwhelming that I struggled to breathe.

In a few hours, Tom was transferred to another hospital that specialized in head injuries. By that time, other members of our family and close friends had arrived at the hospital. There were tears. There were hugs. I'm sure there were prayers. There was disbelief and somewhere in the mix there was also hope.

At one point a little after midnight we were all crammed into a small room at the hospital as we waited to hear further news from the medical staff. My sixteen-year-old son Steve, Tom's brother and best friend, and I were cradled together on the floor as we tried to grab a little sleep which proved to be impossible.

Somewhere around four or five in the morning, a doctor informed us that Tom had no brain activity. Although a ventilator was keeping his heart beating, Tom was clinically dead. Without a functioning brain, there could be no future life as we know it for Tom.

After saying an incredibly difficult goodbye to Tom, his organs were transported to area hospitals to save the lives of other people waiting for organ transplants. While other people lived on with Tom's organs, Tom did not become any of those people.

Was there some form of afterlife for Tom? At times of a tragedy like this, it's easy (and helpful) to think that there is. Perhaps, by the end of this book, you might have a better answer to this question. I'm sure that many family members and some of Tom's friends still think of past happy times with him once in a while. I know that there are moments when I find myself smiling as I recall some of my happy memories with him.

On a side note, when we returned home from the hospital after Tom's tragic accident, it was not very long before his friends filled our house and even our front lawn. In spite of the overwhelming grief that I was experiencing at this time, I couldn't help but hear his friends talk about Tom being "up there" still playing hockey, or telling jokes, or just being with his friends. I never heard anyone talk about Tom being with God. As humans, it's the desire of most people (as explained in an earlier chapter) that life after death is a continuation of the best that we experience here on earth. Could this really be true?

As we consider our lives, it's important to examine the role of our brains. It could be said that we are our brains. As explained in earlier chapters in this book, scientists can elicit a wide range of responses from our brains, including our feelings, our moods, our memories, and our emotions by using electrical stimuli or even by using certain drugs targeted at specific parts of our brain. Any modification or change in our brain (whether through an accident, the use of drugs, or even meditation) can alter our behavior and our personality. Everything about us is stored within our brains. In his book *The Astonishing Hypothesis*, Francis Crick—a physicist and molecular biologist—writes:

> *"You, your joys and your sorrows, your memories and your ambitions, your sense of personal identity and free will, are in fact no more than the behavior of a vast assembly of nerve cells and their associated molecules."*[2]

Could Our Brains Live Forever?

Science and technology in the 21st century may give us the possibility of keeping our brains alive forever. For some, this will be the elusive fountain of youth; for others, this could mean being trapped in a nightmare for eternity. Movies such as the *Matrix* or *Inception* have already portrayed humans living in other dimensions or in a self-imposed world of dreams.

During this century, will science provide a way for us to download our brains into a computer, or find a way for us to keep our brains alive and healthy forever? Will the virtual reality experiences that are currently a part of our entertainment world become a choice for each of us to create our own unique present and future?

Is This the Real Life, or Is It Just Fantasy?

Let's begin by looking at how technology and humans are beginning to merge. In their book *Infinite Reality*, authors Jim Blascovich and Jeremy Bailenson write:

> "The brain doesn't care if an experience is real or virtual. In fact, many people prefer the digital aspects of their lives to physical ones. Imagine you never aged, could shed pounds of cellulite, or put on muscle mass at the touch of a button."[3]

In the book *Ready Player One* (and the movie of the same title), we find a future world where most people live their lives through their avatars in a virtual world. There is no doubt that huge numbers of people already live immersed in digital technology. Go anywhere at any time and you would struggle to find people who are not glued to their cellphone or other digital devices.

Have Video Games Replaced School?

In a pilot study reported in the *International Journal of Mental Health and Addiction*, it was found that 41% of online gamers acknowledged that they use gaming as an escape.[4] It has been reported that 155 million Americans play video games regularly with the average age of gamers being 35 and the average number of years involved in playing games is 13.[5] This would equal the amount of time most students spend from kindergarten to high school graduation. While many people might believe that the video game market is driven by teenage boys, the ESA (Entertainment Software Association) found that women 18 and older represent 33% of the game-playing population while boys aged 18 and younger comprise only 15% of the game-playing population.

Do we really need statistics? Look anywhere and you find people playing video games, and every year the entertainment value of these games is increasing substantially. Consider this quote from Michael D. Gallagher, the president and CEO of the Entertainment Software Association:

> *"Video games are ingrained in our culture. Driven by some of the most innovative minds in the tech sector, our industry's unprecedented leaps in software and hardware engages and inspires our diverse global audience. Our artists and creators continue to push the entertainment envelope, ensuring that our industry will maintain its upward trajectory for years to come."*[6]

Is Virtual Reality More Preferable to Actual Reality?

VR simulations have been with us for a long time even if these simulations weren't called virtual reality. For example, from the earliest days of flying, pilots practiced their skills on simulators before testing their skills in an actual plane. In 1929 the Link Piano and Organ Company of Binghamton, New York, developed a system to simulate actual flying for pilots in training.[7] In looking at the U.S. Apollo program in their endeavor to land on the moon, a great deal of training time for the astronauts was spent in working with simulators. After all, they couldn't just jump into a rocket and fly to the moon to practice.

Although various simulators may have existed in industries such as aviation and the space program for some time, the actual term "virtual reality" is generally attributed to Jaron Lanier who either originated or popularized the term in 1978. Lanier is a scientist who co-founded VPL Research, one of the first companies to sell VR goggles and gloves, in 1985.

Practical Applications of VR

Today, VR simulations are employed in many industries beyond the aviation or space programs. Physicians can practice their surgery skills using VR simulators without risking any harm to patients. A software program called Surgical Theater can create a 3D model of tumors, as an example, and allow a doctor to better prepare for the actual surgery. Each year, between 100,000 to 200,000 patients die from mistakes during operations in America. VR could dramatically reduce these errors. At the Hospital for Sick Children in Toronto, doctors have designed a 3D printed heart which allows a doctor to prepare for surgery on the tiny valves in the hearts of children.[8]

Members of the military use VR to train soldiers for various combat situations. LTC Michael Stinchfield of the Combined Arms Center's Training Innovation Facility at the National Simulation Center stated that with the help of immersive VR experiences, soldiers can be better trained in a more efficient manner.[9] VR can help to train soldiers in a simulated environment without putting anyone at risk. Such training can save time, cut costs, and save lives.

Virtual reality can also be used in the treatment of phobias and to help PTSD patients. In the mid-90's, VR was used to treat PTSD patients who were vets from the Vietnam war. VR was later used to treat patients who experienced PTSD from 911 and also returning sufferers from wars in Afghanistan and Iraq.[10]

Skip Rizzo, Director for Medical Virtual Reality at USC Institute for Creative Technologies, used the graphics engine from a popular combat game called *Full Spectrum Warrior*, to create a virtual Iraq that allowed battle scenes to be recreated for soldiers to help them overcome PTSD. In 2010, Rizzo received the American Psychological Association Award for Outstanding Contribution to the Treatment of Trauma.

Virtual reality can be used in a wide range of engineering and construction projects where a model of the project can be created and tested before proceeding to the actual construction. Price & Myers structural engineer Arthur Coates states,

> *"The greatest potential of Virtual Reality is as a learning tool. Technologies like Microsoft HoloLens, HTC Vive and Oculus Rift allow us to move around an astonishing, high quality projection of the building, learning about the structure at a rate far exceeding traditional no-screen views."*[11]

In 2013 the Ford Motor Company used its Immersion Laboratory in verifying more than 35,000 details of 193 virtual vehicle prototypes.[12] The cost savings and potential safety benefits of such applications is huge.

Virtual reality is even gaining a foothold in sports. For example, NFL quarterback Carson Palmer used VR technology to review his plays. By slipping on a headset which gave him a 360 degree view of the football field, Palmer re-created game situations as though he was standing on the field. At the age of 36, and after returning from what some thought might be a career-ending knee injury, Carson led his team to their best franchise record. He also put up his best career numbers, all of which came after regularly using VR as in integral part of his training. By recognizing game scenarios through the use of VR, athletes are finding that they can react in more positive ways in actual games than they could have through the use of video or other traditional methods.

Virtual Vacations

Another huge application of VR will be virtual vacations where people can relax in the safety and comfort of their homes to take a virtual vacation anywhere in the world. More people die each year from car travel than the current annual deaths from wars. In addition to VR providing a safer way to travel, it would also significantly reduce the carbon footprint which results from our fuel-hungry polluting airplanes.

How many people feel more exhausted than rested at the end of a vacation because of travel hassles such as time delays at airports? VR could eliminate these concerns. Instead of five hours on an airplane to reach your vacation destination (and more hours getting to the airport for your flight and getting to your hotel after your flight), you could be there in minutes, and you could choose who you want to be with and what you want to do without taking out a loan to pay for the vacation.

VR travel guides have already appeared. For example, you can purchase *The Unofficial Tourists' Guide to Second Life*, published by St. Martin's Press where you can obtain advice on the best tours that you might experience in *Second Life*, a leading online virtual program. Imagine mountain hiking on Mt. Everest or scuba diving in Tahiti, or dancing away the night at some pulsating club set on a tropical beach, from the comfort of your own living room.

VIRTUAL REALITY / 177

Be a Gladiator in the Colosseum
The benefits of VR for education are huge. You could take an interactive tour through a remote jungle or travel anywhere in the world without any of the risks. Imagine walking through any museum or city in the world at your leisure with the guide of your choosing. You could even have a teacher who looks and acts in a manner that is most appealing to your style of learning, who could guide you step-by-step through any new learning. And as the technology develops further, you would even be able to journey back in history to experience wonders such as the Roman Colosseum firsthand, as a spectator or as a gladiator.

Google, Facebook, Sony, Amazon, and Samsung
We are beginning to see some huge tech companies invest significantly in VR. In 2014, Facebook purchased Oculus VR for over $2 billion. Samsung Gear provides VR systems to consumers. Google has developed a VR system called Daydream. Sony's PlayStation is taking VR mainstream. It is predicted that more than 110 million VR headsets will be in consumer use by 2020.[13] Amazon has released a video game called Lumberyard which supports VR applications. In addition, Amazon is said to be working on "smart glasses" that could interact with the Alexa environment (which would be mixed reality, rather than solely VR).
What new advances in VR will we see from such powerful companies?

We Can Be Anyone We Choose to Be
In many ways, VR engulfs a user. Experiencing VR can take us out of our normal world and place us into a new world, which when state-of-the-art equipment is used can immerse us in an experience that is as real as the world we have left. It is said that VR creates an almost mystical experience for users known as "presence." VR causes the mind to believe that you're actually present in the VR world.
Considering the huge number of people who play video games on a regular basis, how much more addictive will VR become when we are able to create new identities through our avatars and enter new and exciting worlds where we can be anyone we choose to be?
As Jeremy Bailenson writes in his book *Experience on Demand*:

> *"The psychological effects of VR can be profound and long-lasting. Study after study has shown the experiences that people have in VR have an impact on them. Their behaviors can change, and these changes don't disappear right away. This leads to a conclusion that captures the considerable promise and perils of the medium. VR feels real, and its effects on us resemble the effects of real experiences."*[14]

Do You Prefer Your Dreams or Reality?

Daydreaming, either consciously or subconsciously, is something that everyone experiences. According to a related article in *Psychology Today*:

> *"Some studies have shown that for most of us, our minds wander off somewhere else for about half of our waking hours."*[15]

Michael Corballis in his book *The Wandering Mind: What the Brain Does When You're Not Looking* writes:

> *"Mind-wandering allows us to inhabit the minds of others, increasing empathy and social understanding. Through mind-wandering, we invent, tell stories, expand our mental horizons. Mind-wandering underwrites creativity, whether as a Wordsworth wandering lonely as a cloud, or an Einstein imagining himself traveling on a beam of light."*[16]

In a sense, we have all already created virtual worlds that we retreat to when we're bored or we simply lose focus. We all daydream. Virtual reality will allow people to enhance or better control this experience, and perhaps more significantly it will allow people to enter virtual worlds where they can actually interact with others who are seeking similar experiences.

Have a Second Life

The virtual game *Second Life*, launched in June of 2003, boasts on its Facebook page that it is "the leading 3D virtual world where you can be whomever you'd like and create anything you can imagine." Estimates of *Second Life* users suggest many million participants generating multi-millions in revenue. With such financial benefits for the originators, reality games will continue to expand providing worlds that can be created by their participants.

Games such as *Second Life* allow users to create a new way of living where people can select an avatar that mirrors whoever they want to be. These games provide the opportunity to have complete control over how events transpire. It is as though users have become God. They can choose whatever is going to happen next in their virtual lives and they can choose the outcome. Making such experiences even more realistic (or addictive, depending on how you look at it) is that users, through their avatars, can interact with other users. The restraints, that apply to normal everyday living, are all gone when users immerse themselves in VR experiences.

In competition with *Second Life* there are already a host of similar games such as *Active Worlds*, *Twinity* (where you can become a Twinizen), *IMVU* (a virtual universe where your avatar can meet and chat with others online), *Entropia*, *Kaneva*, *there.com*, *onverse*, and likely many more by the time this book is published.

Observer or Participant?

Some readers might dismiss such platforms and games as a form of entertainment, not much different than escaping in TV shows, or other video games. The difference, and this is important, is that when we watch TV or play most video games we know we are watching TV or playing a game. Yes, a well-directed TV show with compelling acting and a well written story can briefly transform us to another world, but we never actually interact with the characters in that world. Virtual reality changes this. As has been previously stated in this chapter, participants in VR often have trouble differentiating what is happening from that of a real-life experience. In addition, VR offers the possibility of actually interacting with others. When we watch TV, in spite of any suspense, we are passive viewers. When we engage in virtual reality, we become active participants.

Escaping Life's Problems Through VR

How might virtual reality change the lives of individuals and impact our world?

Second Life quickly became popular with physically disabled people because it allowed them, through their avatars, to lead physically normal lives. *Second Life* allowed these people to run and jump, to be sports heroes, to live the kind of life they often imagined. Virtual Reality programs like *Second Life* also allow people who are socially awkward or who perceive themselves to be unattractive to create new identities where they can become movie-star beautiful living the lives of celebrities. And for the millions of people who work in boring jobs, *Second Life* provides an opportunity to live a thrilling existence where they can be anyone they want to be and do whatever they want without having to be rich.

Consider the multitude of problems that many people face including obesity, money problems, relationship problems, family problems, and problems at work. With a keystroke, these problems can all disappear in virtual reality. The easy-to-achieve solutions are so realistic that VR may become the "soma" that Aldous Huxley described in his novel *Brave New World*. Huxley wrote that soma had all the advantages of Christianity and alcohol, with none of their defects.

Could VR become soma for the world?

Being present in a virtual world can take a person away from whatever pain, whether physical or emotional, that they might be suffering in the real world.

Virtual reality is accessible to millions. It is easy to use. It's cheap in comparison to other forms of entertainment. It's available on-demand, and it's available to use almost anywhere a person happens to be.

Connecting the World

VR can help connect people across the planet, to better understand the hopes and the concerns of people we would not normally come into contact. As an example, Chris Milk, working with the United Nations and Samsung, created *Clouds over Sidra*, a VR docu-

VIRTUAL REALITY / 181

mentary. According to the UN, this documentary doubled the number of people who might have been expected to donate money to the cause.[17]

Could VR Eliminate Pain and Suffering?

Virtual reality can also have a direct application related to dealing with pain and suffering, a topic we explored in previous chapters.

Consider a person who is suffering pain from an accident or illness. Virtual reality can provide temporary relief from this suffering by transporting the person to another world where they are no longer confined to a bed or restricted by their health condition. It's even possible that in the same way that VR is used in many different training situations (such as pilots or athletes) that virtual reality might help people overcome illnesses and injuries as the brain visualizes healing.

VR has been shown to be effective in the treatment of anxiety and other psychiatric disorders. In the 2017 *Harvard Review of Psychiatry*, we read:

> *"VR aims to parallel reality and create a world that is both immersive and interactive. Users fully experience VR when they believe that the paradigm accurately simulates the real-world experience that it attempts to recreate. In all, VR is potentially a powerful tool for the psychiatric community because the user experience can be consistently replicated, tested, and modified within a safe environment."*[18]

Consider those who live in constant pain from an accident or illness. In a June 2014 article from the *U.S. National Library of Medicine*, National Institutes of Health, we read:

> *"VR has been found to be effective in reducing reported pain and distress in patients undergoing burn wound care, chemotherapy, dental procedures, venipuncture, and prolonged hospital visits. Chronic pain patients demonstrated significant relief in subjective*

> *ratings of pain that corresponded to objective measurements in peripheral, noninvasive physiological measures."*[19]

For some who may be suffering from an illness or accident, or who may have decided that life is no longer worth living, VR may provide an alternate world where suffering, pain, and even depression no longer exist.

VR and Aging

With a growing senior population, VR may help seniors better deal with some of the common issues of getting older.

Common complaints of the elderly are that they don't want to suffer, and they don't want to become a burden on their family (and they don't want to be alone).

The elderly face loneliness in unprecedented numbers. VR could help with these concerns. It is an unfortunate fact for many elderly people that family can't always be physically present to help them deal with their final days. VR could provide a world for some seniors where they could interact with others, possibly even the avatars of other family members who live too far away to otherwise visit. Unlike Skype or FaceTime, VR could be available on demand.

VR could also help the elderly engage in activities that they loved to participate in throughout their lives but now due to age (and confinement due to either physical or mental deterioration) are unable to do so.

Rendever is a company that specializing in using VR for the elderly. CEO Dennis Lally states:

> *"They can go to a Maui beach and watch the waves come in for 30 minutes, or swim with a whale in the ocean. They could sit in the front row of a concert that they wouldn't otherwise be able to attend. We also provide educational stuff, like historical tours or architectural exhibits."*[20]

Researches are exploring how VR might improve the lives of seniors by helping them to deal with dementia, depression, and anxi-

VIRTUAL REALITY / 183

ety. Lora Appel, a researcher with OpenLab at Toronto General Hospital, believes that VR can improve the quality of life for seniors with dementia to allow them to go virtually outdoors without facing the dangers they might otherwise encounter due to their changing mental capabilities.[21]

Virtual Immortality
Back in Chapter 5, we looked at how traditions from thousands of years ago have shaped our beliefs related to life after death. VR is providing a new way to think about eternal life.

In addition to improving the quality of life for some who may be confined to a bed, VR may also present the possibility of virtual immortality. VR offers alternate worlds to users. VR offers a chance to create one's own heaven. And VR doesn't require a body to participate. The only requirement is a brain. It is quite possible in the near future that technology will advance to the point where a brain can be kept alive artificially by mechanical means even if the rest of the body has perished. It is also possible that technology will advance to the point where a person's brain can be downloaded into a computer where the person will be able to experience virtual immortality.

Futurist Ian Pearson believes that we will achieve some form of virtual immortality by downloading our brain into a computer by the year 2050 if you're rich; 2075 if you are poorer.[22] A recent article in the *London Express* commented on Pearson's predictions:

> *"Immortality has been regarded as mythology and science fiction for years but now human beings are close to defying death due to several major scientific breakthroughs which will give humans a plethora of choices on how to live forever by the year 2050, according to a top futurologist."*[23]

Others are less optimistic, but there is no doubt that there are scientists who are actively exploring the possibility of virtual immortality. Many of the technological advances that we currently enjoy would have been unheard of 50 years ago, so who knows what the next 50 years will bring?

Creating Digital Versions of Ourselves

The National Science Foundation has awarded a half-million-dollar grant to the universities of Central Florida at Orlando and Illinois at Chicago to explore how digital versions of real people might be created.[24]

The "2045 Initiative" is a nonprofit organization, founded by Russian entrepreneur Dmitry Itskov, that is researching life extension by creating technologies to enable the transfer of an individual's personality to a non-biological carrier, thus extending existence to the point of immortality.[25]

Grieving the Loss of a Loved One through VR

When people die, grief impacts the survivors. Regardless of what happens to a person who has died (whether heaven exists, or whether there is nothing at all beyond death), it is the family members and friends who grieve the loss of the person.

A future option that some may choose is to set up an avatar in a virtual world whose voice and personality is a replica of the deceased person. This would permit a family member to continue to interact with a loved one in a virtual world, and while it is not helping the deceased, it might help family members and friends to have some form of continuation with a loved one. Whether this may prove to help in the grieving process remains to be seen. It could be argued that it could prolong the grieving process indefinitely, although some might also argue that it might serve to shorten the grieving process through a VR interaction with a living avatar of a deceased person.

The Silicon Valley-startup Eterni.me is attempting to help people by preserving their most important thoughts and experiences in a VR system that would allow family members and friends to communicate with a deceased person's avatar. Could this help people better deal with the loss of a loved one? There is no evidence at this point in time to appropriately answer this question.

There have already been movies such as *Forrest Gump* where famous people who have previously died were recreated on the big screen. The late actor Paul Walker, famous for his role in the *Fast and Furious* series, was "reanimated" after his death to reappear in *Furious 7*. Famous entertainers, such as Amy Winehouse and Roy

Orbinson, who have died have reappeared as holograms at concerts, in award shows, and even in commercials. The Japanese singer Hatsune Miku regularly sells out stadiums and arenas for her concerts in many parts of the world, but Miku is not even real—she is a hologram. Miku is an invention of virtual reality, never having been a real person.

From what we have seen in movies, future avatars could become virtually indistinguishable from the person they are representing. They could also be the fantasy perfect version of who a person wants to be.

Cyber-Immortality

William Sims Bainbridge, program director of the National Science Foundation, has sent his avatar and DNA to be stored at the International Space Station. Bainbridge is a "cyber-immortality" enthusiast, and as such he has developed projects and software related to this. One free software program he has developed is titled *The Year 2100*. This program is intended to store a person's personality including values, hopes, beliefs, etc. that might someday be reanimated through an avatar or even a new biological body.

Obviously one of the real obstacles in the path of virtual immortality relates to consciousness. Robert Lawrence Kuhn, the creator, writer and host of "Closer to Truth", a public television series, questions whether human consciousness can ever be duplicated in a computer. Kuhn states:

> *"Unless humanlike inner awareness can be wholly recreated, even synthesized, by physical manipulations alone, uploading one's neural patterns and pathways, however complete, could never preserve the original, first-person mental self (the private "I"), and virtual immortality would be impossible."*[26]

VR Dangers

VR could create problems. One of the more obvious concerns with using VR is that a person immersed in VR may attempt to jump or react in some physical way to whatever they are actually doing in their virtual world. Without a person protecting a user from danger, a VR user could fall or run into a wall, or even crash through a window.

Another concern with VR relates to the growing number of people who suffer from nearsightedness. Nearsightedness rose in the American population from only 25% of the population in the 1970s to over 40% by 2000. It is thought that our constantly looking at tablets and cell phones has contributed to this problem.[27] The use of VR could increase this concern. Along the same lines wearing a virtual reality headset could impair hearing over time. Given that VR is used to train soldiers to carry out combat missions without morally or emotionally reacting, is it also possible that graphic violence in VR could lead a person to engage in such behavior in real life?

The use of VR can cause cybersickness which is similar to motion sickness. More serious is the question as to how VR might impact us neurologically. Gaming (non-VR) has been known to cause isolation and reclusive behavior. How will a more immersive system like VR affect some people, especially those who may already be experiencing mental health concerns?

VR is Here to Stay

Regardless of any physical or psychological dangers that might result from VR, it is here to stay. Over the past few decades there have been concerns voiced about the negative impact of people burying themselves into surfing on the Internet or being glued to their cellphones. In spite of any warnings that might have been delivered, stop anywhere in a public place and as you observe the people around you, you will quickly conclude that our love affair with digital devices such as the cell phone is increasing, rather than diminishing. As VR becomes more sophisticated and lifelike, its usage will undoubtedly increase.

While we may very well see virtual immortality or some variations of this before this century ends, by the time this happens we may be living in a very different world, a world where the lines of what is real and what is artificial become blurred.

Are We Already Living in a Virtual World?

There are those who believe we are already living in a virtual

VIRTUAL REALITY / 187

world. As stated earlier in this book, such thinking would have been rare 50 years ago, primarily because the terminology and vocabulary associated with VR didn't even exist, but now that it does we have a whole new way of explaining our reason for being here. We will explore such thinking further in Chapter 11.

- 9 -
Three Degrees Fahrenheit

In 1953, the dystopian novel *Fahrenheit 451*, written by Ray Bradbury, was published. This book presents a futuristic society in which books are banned and firemen burn any that are found. The title of the book—Fahrenheit 451—is the temperature at which the pages in a book catch fire. In the 21st century, we find ourselves looking at the significance of temperature in a different light

A few years ago, my family and I went away on a short vacation on a weekend late in May. At the time we had a 100-gallon saltwater aquarium set up in our family room. The aquarium was home to a few beautiful fish along with an assortment of striking corals (grown from frags cultivated at coral farms rather than being harvested from the wild). During the few days that we were away, an unexpected heat wave struck our city. Normally we would have had our air-conditioning on during such weather, but in this instance the forecast had not predicted high temperatures, so we left home without turning it on.

When we returned home, we sadly discovered that everything in our aquarium had perished. The fish were dead. The corals were dead. In looking at the thermometer on the aquarium, the temperature was 3°F higher than normal. Three degrees! It only took a 3°F difference to cause a tipping point for life in the aquarium.

Death of the Great Barrier Reef?
A recent report on CNN stated, "Marine heat waves caused by global warming are killing off the corals of Australia's Great Barrier Reef, the world's largest reef system."[1] The report described how heat waves in 2016 and 2017 had destroyed half of the corals on the Great Barrier Reef. Researchers for this study said that a rise of more than 2°C (or 3.6°F) would doom the Great Barrier Reef (as

well as other reefs around the world). With reefs providing a living home to countless varieties of fish, it is estimated that more than a billion people worldwide derive their food from reef areas.

What is the Tipping Point for Life on Our Planet?

In considering the impact of global warming, we might ask what is the tipping point for life as we know it on our planet? In 1975, the economist William Nordhaus suggested that 3.6°F was likely the answer to this question. In 1996 the European Council recommended that 3.6°F should be the red line of the United Nations for global warming. In the year 2000, countries attending a climate change conference in Mexico committed to preventing this 3.6°F increase from occurring.

While there are some experts who question the 3.6°F threshold, and would argue that this figure is too low, even those who disagree with the 3.6°F threshold would rarely argue that this number could ever double to 7.2°F without catastrophic effects.

Putting this into a different perspective, a 5°F drop 20,000 years ago buried a large part of North America under a huge mass of ice. What catastrophic impact might a 5°F rise in temperature have for us in the 21st century?

Global Warming and the Return of Jesus

In 2016, 178 nations signed the Paris Agreement on Climate Change to go a step further in setting a new maximum worldwide temperature rise of 1.5°C (2.7°F) as the threshold to stay within. In 2018 the U.S. government withdrew from this agreement negatively impacting its future success as the U.S. is the second largest emitter of harmful emissions (and it should be noted that the American government who made this choice was voted into power by its base whose greatest membership is evangelical Christians—need I say more about how religion continues to cause problems in the 21st century?).

As it has been stated elsewhere in this book, some members of conservative religions, such as the evangelicals in the U.S., believe that God is in control (when this belief suits their needs), so there are times when they don't feel a need to be responsible for problems such as climate change because God will solve this problem

for them.
U.S. Congressman Tim Walberg told his constituents:

"As a Christian, I believe that there is a creator in God who is much bigger than us, and I'm confident that, if there's a real problem [in regards to climate change], He can take care of it.²

Added to this abandonment of responsibility because God will look after things for us, the belief that Jesus will be returning soon dismisses many fundamentalist Christians from being concerned about climate change. Randall Balmer, professor of religion at Dartmouth College, stated (in reference to Scott Pruitt—the current Head of the U.S. Environmental Protection Agency):

"If you believe Jesus is coming back at any moment, why bother with social reforms, why bother with environmental protection?"³

Unfortunately, a reading of any historical records related to Christianity shows that many Christians have been expecting the return of Jesus at any moment for the past 2,000 years. According to historian Charles Freeman:

"Early Christians expected Jesus to return within a generation of his death and the non-occurrence of the second coming surprised the early Christian communities."⁴

Amazingly most of the Christians today who believe that Jesus will be returning at any moment still have jobs, houses, make plans for future events, and so on. One might ask them, "If Jesus is going to return at any moment, why bother with any sort of work obligations or future planning?" Isn't it a little hypocritical to ignore climate change because Jesus is about to return, but have no problem enjoying all the comforts that technology and science otherwise provides?

Discounting Science
Another reason why many evangelical Christians disbelieve cli-

mate change is there is a general tendency among this group to discount science in any form. After all, many of these Christians are the same people who still don't believe in evolution in spite of overwhelming scientific evidence to support evolution as a fact.

The resistance to evolution or climate change from conservative Christians is a reminder that back in 1610 the Catholic Church's refused to accept Galileo's thoughts that the earth was not the center of the universe. It took more than 200 years (1822 to be exact) before the Catholic Church acknowledged this scientific fact. If we wait more than 200 years for conservative Christians to acknowledge the problem of climate change, the human species may be extinct.

I would imagine though that the same people who mock the science of evolution or climate change readily accept the science of fighting diseases with appropriate medicine (although there are a few religions such as Christian Science who reject some aspects of modern medical science and believe that prayer is cure for any illness).[5]

You might wonder why it could possibly be a problem if evangelical Christians don't believe in climate change. The answer is fairly simple. Evangelical Christians form a large part of the base of supporters for the Republican party in the U.S. When the Republican party is in power, any efforts to address the problem of climate change will be ignored because the Republican leadership needs their evangelical Christian base to help them get elected. This can result in one of the most powerful countries in the world resisting any attempts to deal with climate change.

Of course, it would be inaccurate to try to blame the failure of addressing climate change on just evangelical Christians. There are millions of people, dare we say billions, across our globe who have no interest (or knowledge) in the problem of climate change. For some people who are struggling to meet their basic needs each day, climate change is not a relevant concern. In addition, there are those who think that climate change is caused by pollution from big companies, and as such they don't think there is anything they can do as individuals. And finally, there are some that think that the problem simply cannot be solved.

THREE DEGREES FAHRENHEIT / 193

The Major Cause of Climate Change
The earth has experienced severe forms of climate change as a result of natural causes before (think ice ages that significantly altered life on the planet many years ago), but this time the impending climate change could be the result of human activity (and it might be preventable if we were to alter our harmful emissions into the atmosphere and stopped destroying natural vegetation on our planet).[6]

From the National Research Council (2010) we read:

> *"Some scientific conclusions or theories have been so thoroughly examined and tested, and supported by so many independent observations and results that the likelihood of subsequently being found to be wrong is vanishingly small. Such conclusions and theories are then regarded as settled facts. This is the case for the conclusions that the Earth system is warming and that much of this warming is very likely due to human activities."*[7]

On Jan. 25, 1984, in his State of the Union Address, President Ronald Reagan said, "Preservation of our environment is not a liberal or conservative challenge: it's common sense." It would appear that the current ruling party of the U.S. and their evangelical Christian base have abandoned any notion of common sense. In a *New York Times* article from August of 2013, four U.S. Republican leaders of the Environmental Protection Agency who served four different U.S. presidents stated:

> *"We served Republican presidents, but we have a message that transcends political affiliation: the United States must move now on substantive steps to curb climate change, at home and internationally. There is no longer any credible debate about the basic facts: our world continues to warm, with the last decade the hottest on record, and the deep ocean warming faster than the earth's atmosphere. Sea level is rising. Arctic ice is melting years faster than projected. Climate change puts all our progress and successes at risk. If we could articulate one*

framework for successful governance, perhaps it should be this: When confronted with a problem, deal with it. Look at the facts, cut through the extraneous, devise a workable solution and get it done."[8]

Fact or Hype?

Global warming is based on scientific fact. While there are some scientists who dispute the impact that global warming will have on our planet, there are few who actually deny that it is happening. A group of 1,300 independent scientific experts from countries around the world concluded (in its Fifth Assessment Report):

> *"There's a more than 95% probability that human activities over the past 50 years have warmed our planet. The industrial activities that our modern civilization depends upon have raised atmospheric carbon dioxide levels from 280 parts per million to 400 parts per million in the last 150 years. There's a better than 95% probability that human-produced greenhouse gases such as carbon dioxide, methane and nitrous oxide have caused much of the observed increases in Earth's temperatures over the past 50 years."*[9]

Global warming is largely the result of carbon dioxide (from the burning of fossil fuels such as coal and gas) and methane (from decomposition of wastes in landfills as one example) being released into our atmosphere and the destruction of natural vegetation that could help to convert harmful carbon dioxide into oxygen. While there has always been varying degrees of carbon dioxide in our atmosphere, current studies tell us that the amount of carbon dioxide in our atmosphere is 40% higher than it has been in the past 800,000 years, and it is expected to double before the end of this century.[10]

To measure the amount of carbon dioxide from the past, scientists bore holes up to 3.2 km (2 miles) in depth into arctic ice enabling them to measure the presence of this gas in air bubbles from many hundreds of thousands of years ago. From the data that has been measured, there is a sudden dramatic rise in carbon dioxide in the atmosphere beginning in 1950, a rise that was greater than any of the past 400,000 thousand years.[11]

When heat from the sun (and factories, houses, and many other human activities) reach the surface of the earth, some of the heat is radiated back away from the earth. When a gas like carbon dioxide builds up in our atmosphere it acts like a blanket preventing heat from escaping from the earth and as a result our planet warms.

Al Gore, former U.S. vice-president and award-winning environmentalist, writes:

> *"We're still treating the atmosphere as an open sewer. We're putting 110 million tons every day of man-made heat trapping pollution into the sky. And it lingers there for a long time. The cumulative amount now traps as much extra heat as would be released by 400,000 Hiroshima-class bombs exploding every day."*[12]

Proof that Climate Change is Happening

Whatever the temperature threshold is that would trigger a tipping point where life would perish on a massive scale throughout our planet, we are already moving towards it. In a recent article on CNN News, Paul Hockenos writes:

> *"This summer's sizzling temperatures, savage droughts, raging wildfires, floods and acute water shortages—from Japan to the Arctic Circle, California to Greece—are surely evidence beyond any reasonable doubt that the climate crisis is upon us now. This is the new normal—until it gets worse. We, the entire global community, the residents of this planet, must finally grasp the urgency at hand and undertake dramatic, meaningful measures. It's the mainstream opinion among serious scientists worldwide: climate change, unchecked, will eventually wipe out our race—and man-made greenhouse gases are still rising."*[13]

In looking at information from NASA, 17 of the 18 warmest

years on our planet have occurred since 2001 (over a 136-year period of record keeping). In 2016 the global average temperature was 0.94°C (1.69°F) above the 20th century average.[14] While a rise of 1°C might not seem like much, what appears to be a slight rise in temperature could stimulate other factors that could increase the rate of global warming.

A significant amount of ice is continually melting in the Artic, Greenland, and West Antarctic exposing dark ocean waters that absorb more heat in comparison to the ice that reflected much of the heat. While the complete collapse of the Greenland and West Antarctic Ice Sheets could take hundreds of years, the end result could be a rise of up to 20 feet in our oceans, devastating major coastal cities and disrupting food supplies around the world. Even a rise of 3 feet could cause horrendous problems when storms strike coastal areas.

Adrienne White, a glaciologist from the University of Ottawa, who studies glaciers in Canada's High Arctic found that between 2000 and 2016, of 1,773 glaciers studied, 1,353 shrank significantly.[15] As glaciers melt, the oceans will get warmer, not only threatening life in the oceans, but contributing to the intensity of storms and hurricanes that are fueled by warmer waters.

In addition to the problems caused by melting ice, warmer temperatures are causing the permafrost in high latitudes to melt. The permafrost layers in the ground in these locations have trapped carbon dioxide and methane for thousands of years. As the permafrost melts, these harmful gases will be released into the atmosphere causing further global warming. In addition, methane is stored harmlessly in the floor of the ocean. As ocean temperatures rise, this could cause a massive release of this methane gas, thus speeding up global warming.

The Impact of Global Warming

There are a wide range of potential devastating consequences of climate change. Droughts caused by global warming kill massive amounts of vegetation. As vegetations dies, all animals (including us) will lose a major source of food, and the death of our vegetation will increase the amount of carbon in the atmosphere (plants convert carbon dioxide to oxygen so as our vegetation disappears there will be more carbon dioxide entering our atmosphere). In addition, vegetation helps to cool our planet. As vegetation decreases due to global warming, this will increase the warming problem.

Where I live in Canada, this past summer we experienced 10 consecutive days of very hot, dry weather. Walking around the neighborhood during this time, there were large amounts of leaves dropping from trees. In countries and areas where droughts are prolonged, vegetation will die from a lack of water and from possible wildfires which will lead to more harmful carbon dioxide in our atmosphere.

As oceans warm, this can fuel more intense storms and hurricanes resulting in serious flooding in coastal areas as well as killing an enormous variety of sea life that provides food for billions of people.

It should be noted though that not every natural disaster such as hurricanes, typhoons, earthquakes, and volcanic eruptions is caused by global warming. In looking at the impact of global warming, and the impact could be huge, we need to be careful that climate change doesn't suddenly become the catchword to explain every disaster. Natural disasters have been with us since the beginning of time, long before global warming, so it would be wrong to blame every natural disaster around the world on climate change.

As I write this chapter, there have been endless articles in the news related to the growing threat of climate change. An article in the *Washington Post* stated:

> *"An intense heat dome has swelled over Scandinavia, pushing temperatures more than 10 degrees above normal and spurring some of the region's hottest weather ever recorded. The same news organization (Sweden's national weather agency) reported the high temperatures have intensified a 'historic wildfire outbreak.'"*[16]

Massive fires are currently burning in California and other parts of the U.S., Canada and Greece. An article from this morning's newspaper was headlined *No mystery as climate change worsens heat, fires.* The article stated:

> "Heat waves are setting all-time records across the globe, again. It's all part of summer—but it's been made worse by human-caused climate change, scientists say. So far this month, at least 118 of these all-time heat records have been set or tied across the globe according to the National Oceanic and Atmospheric Administration."[17]

Will We Destroy Much of the Life on Our Planet?

Dr. Tony Juniper, a leading environmentalist, in his book *What's Really Happening To Our Planet?* writes:

> "Earth's atmosphere now has a higher concentration of greenhouse gases than at any time for at least 800,000 years. This is already causing climate change, leading to more extreme conditions, increased economic costs, and major humanitarian impacts. The scale of ecosystem degradation means that a mass extinction of animals and plants is gathering momentum. This could soon lead to the greatest loss of diversity since the dinosaurs were wiped out 65 million years ago."[18]

For the first time since humans inhabited this planet, we have the potential to destroy much of the life on it, whether the destruction comes from global warming, a nuclear war, some other weapon of mass destruction, or some unforeseen biological plague. In the front jacket of Elizabeth Kolbert's book *The Sixth Extinction*, we read:

> "Over the past half billion years, there have been five major mass extinctions, where the diversity of life on earth suddenly and dramatically contracted. Scientists are currently monitoring the sixth extinction predicted to be the most devastating since the asteroid impact that wiped out the dinosaurs. This time around, the cataclysm is us."[19]

The End of Humans

Whenever people talk about a problem such as global warming, the discussion tends to focus on the impact on nature without fully acknowledging the direct impact on humankind. Part of this problem stems from the significant number of people who do not accept global warming or who simply don't care. As discussed elsewhere in this book, if you are a conservative Christian and you believe that Jesus is about to return, there may be no incentive for you to care about our planet because Jesus is going to create some fantastic new world here on earth. And yes of course there are some Christians who passionately care more about the health of our planet than the imminent return of Christ. And, yes of course there are some people who are not Christians who are also apathetic about anything to do with climate change.

Many people believe that humans are special, that God has made us superior to all other creatures on our planet. These people inherently believe that while global warming might contribute to the death of coral reefs and some other forms of life, it couldn't possibly spell the demise of humans. There are a significant number of people who can't imagine this world carrying on without humankind. Such thinking demonstrates a profound lack of understanding of the evolution of life on our planet.

A recent article in the *New York Times* stated:

> *"Extreme heat can kill, as it did by the dozens in Pakistan in May, 2018. But as many of South Asia's already scorching cities get hotter, scientists and economists are warning of a quieter, more far-reaching danger. Extreme heat is devastating the health and livelihoods of tens of millions more. If global greenhouse gas emissions continue at their current pace, they say, heat and humidity could become unbearable, especially for the poor."*[20]

A recent report published in the American Proceedings of the National Academy of Sciences said:

> *"If the threshold—a theoretical point of no return—is crossed, this would lead to much higher global average temperature than any interglacial in the past 1.2 million years and to sea levels significantly higher than*

at any time in the Holocene (referring to the geological age which began at the end of the last ice age, around12,000 years ago)."²¹

In addition to global warming, there are other possible major concerns for the future of humans. Population growth will be a problem as it taxes our resources and could contribute to greater pollution thus increasing global warming. In the early 1800s, human population reached 1 billion people. It is projected that by 2050 we will reach 9 billion. As our population continues to increase and droughts around the world result from global warming, we will face greater water problems in the future. Although our earth is covered with 71% water, only 2.5% of this is freshwater (and almost 70% of the freshwater is stored in glaciers and ice, some of which is melting and flowing back into the ocean where it becomes useless to us).

It is currently estimated that our earth is a little over 4.5 billion years old. Considering that humans, in our current form, might have only existed for a few hundred thousand years, and taking into account our most direct ancestors who might have inhabited the planet for a few million years, we are nothing but a small blip on the timeline history of our planet. As Bill Nye writes:

> *"A great number of people in many parts of the world—even in well-educated parts of the developed world—are resistant or hostile to the idea of evolution. Even in places like Pennsylvania and Kentucky, here in the United States, the whole idea of evolution is overwhelming, confusing, frightening, and even threatening to many individuals. I can understand why. It's an enormous process, unfolding over times that dwarf a human lifespan—across billions of years and in every part of the world. And it's profoundly humbling. As I learned more about evolution, I realized that from nature's point of view, you and I ain't such a big deal. Humans are just another species on this planet trying to make a go of it, trying to pass our genes into the future, just like chrysanthemums, muskrats, sea jellies, poison ivy,...and bumblebees."²²*

THREE DEGREES FAHRENHEIT / 201

Award-winning author Matt Ridley writes:

"But if there is one dominant myth about the world, one huge mistake we all make, one blind spot, it is that we all go around assuming the world is much more of a planned place than it is."[23]

If the Dinosaurs Could Vanish...

Earth existed with a wide array of life for billions of years before the evolution of humans, and quite frankly our planet will continue even if humans, as a species, become extinct. In considering natural selection, which has been explored widely in this book, there is absolutely no reason why humankind should believe that we are immune to disasters of any sort that could eradicate us. If the dinosaurs could vanish, then so can humans. And it might not take a visible dramatic catastrophe to terminate our place on this planet.

One afternoon a few months ago in the spring, my wife and I were walking through a forest where there were several swampy areas. I paused often, looking in the still waters beside the path where we walking. When my wife asked what I was looking for, I explained I was looking for frog eggs. As a kid, I loved searching streams and ponds for the jelly mass of eggs from frogs and salamanders. I would often take some of the eggs home and watch the development of the eggs into tadpoles and then into frogs before taking them back to the streams and ponds where I found the eggs.

The last time I found any frog eggs on a walk dates back at least three decades. Recently on a family canoe trip in a natural protected wilderness area, I told stories to my daughters that a few decades ago, the streams where we were paddling were inhabited by endless huge bullfrogs. But no more! Where did all the frogs go?

Frogs have been disappearing along with many other amphibians at an alarming rate over the past 40 years. It is currently thought that a chyrid fungus, known as Batrachochytrium dendrobatidis, is the silent killer for frogs.[24] Given that frogs and other amphibians have a complex immune system, not too dissimilar to that of humans, is it possible that such a killer, or some super-bacterial infection, could take root in humans and start us on a slow death march to extinction?

Along the same lines, 10 years ago scientists raised the alarm that bees were dying in massive numbers across the planet. This phenomenon, known as Colony Collapse Disorder, was attributed to a number of factors including pollution, pesticides and climate change. Bees have survived on our planet for millions of years, much longer than humans. One recent article stated that bees have been present on our planet for more than 130 million years.[25] Bees even survived whatever killed off the dinosaurs (who became extinct about 65 million years ago). It is estimated that close to 90% of land plants (including fruits and vegetables) depend on insects such as bees to pollinate them.

In 2014, the U.S. government announced that it was spending $3 million dollars to protect honey bees.[26] If 21st century environmental concerns have started to eradicate creatures that have survived for millions of years, is it possible that these same concerns may begin to take a toll on humans?

Declining Sperm Counts

While frogs and other amphibians have been quietly disappearing over the past 40 years, we find that during this same period, the sperm counts in men from America, Europe, Australia and New Zealand have declined 59.3%.[27] Researchers involved in this study also concluded that this rate of decline is not slowing.

In the same manner that pesticides and chemicals (and stress) are impacting bees, humans—in this case, males—are also being impacted. What would happen if this trend continued and huge numbers of males across the planet became infertile? It seems like the fabricated plot of a dystopian horror movie, but the implications could become incredibly disastrous.

Patrick Bateson, professor emeritus of ethology at the University of Cambridge, writes:

> *"Another darker thought is that the human population might be curbed by its own stupidity and cupidity. I am not now thinking of conflict but of the way in which endocrine disruptors are poured unchecked into the environment. Suddenly males might be feminized by the countless number of artificial products that*

simulate the action of female hormones—sufficiently so that reproduction becomes impossible."[28]

Climate Change and the Increase of Diseases

We also find that climate change will expose humans to a greater variety of diseases. For examples, mosquitos carrying a disease such as Zika were confined to hot counties that bordered the equator until recently. As the northern hemisphere continues to experience warmer weather, mosquitos carrying the Zika virus are also migrating further north. Lyme disease caused by a certain species of ticks is also migrating further north.

Along with problems with disease-carrying insects, climate change (along with the overuse of antibiotics) is increasing the presence of super bugs, bacteria infections that resist treatment. A recent article in the Canadian Press titled *Dangers of infectious superbugs likened to climate change: a slow and moving tsunami*, stated:

> *"Unless urgent action is taken, experts warn that by 2050, the annual death toll will soar to 10 million worldwide, dwarfing cancer."*[29]

In his book *Homo Deus*, international bestselling author Yuval Noah Harari writes:

> *"Homo sapiens has rewritten the rules of the game. This single ape species has managed within 70,000 years to change the global ecosystem in radical and unprecedented ways. Our impact is already on a par with that of the ice ages and tectonic movements. Within a century, our impact may surpass that of the asteroid that killed off the dinosaurs 65 million years ago."*[30]

For the most part, we take life for granted. In many ways religion has done this to us. With the promise that God is always with us and with the hope of eternal life in heaven, it's far too easy for some to avoid any sense of responsibility for what is happening on our planet in the present. For many people of various religious faiths, the emphasis is on worshipping God and preparing for the hereafter, instead of taking responsibility for what is happening right now on our planet.

In addition, it appears to be human nature for most people to avoid dealing with problems that appear to be far off in the future. When it comes to climate change, even though we are beginning to experience some effects of the concern, the main problem—if it ever happens—is seen to be off in the distant future. World leaders rarely, if ever, plan for the distant future so why should the average person?

Creation Repeated

It is a scientific reality that carbon dioxide emissions contribute to global warming. Recent studies, reported in the *Washington Post*, looked at the impact of a huge meteor that crashed into our planet some 65 million years ago, killing more than 75% of life on our planet including the dinosaurs.[31]

The studies found that the carbon released into the atmosphere from this asteroid resulted in a 5°C warming in the seas and that this impact lasted for 100,000 years. In other words, unless we take action to halt global warming, once we reach the tipping point where life begins to perish, the impact will continue for tens of thousands of years.

Let's return back to my dead aquarium. After removing the dead coral and lifeless fish from the aquarium, I drained about 95% of the water from the tank because it was foul. I left about an inch of water covering the fine sand on the bottom of the aquarium to help block the smell from the decay in the sand. I covered the aquarium and decided to wait a few weeks before transporting it to our basement where the coolness there would prevent a terrible accident like this from occurring again.

A little over a week after removing the dead coral and fish from the aquarium along with most of the water, I noticed small trails in the sand on the bottom of the aquarium. Upon closer inspection I discovered sea worms burrowing in the sand. Although most of the visible life in the aquarium had died after a 3°F rise in temperature, the worms living in the sand remained healthy.

If we ever manage to exterminate our own species or destroy

most of the visible life on our planet, in the depths of the ocean and hidden below the surface of the earth, small creatures will survive. And in the same manner that life began billions of years ago from single-celled organisms, life will begin its slow methodical natural selection journey once again.

If this ever happens, millions, or even billions, of years from now our planet will once again be home to a tremendous diversity of life. Eventually natural selection might result in some creatures evolving to have the mental capacity of humans today. Whether this new species will look like us remains to be seen, but whenever any future creature approaches our intellectual ability, they too will invent new gods, new philosophies, new religions, and likely a self-centered egotistic approach to their place on the planet.

Consider the warning of Richard Preston in his bestselling book *The Hot Zone*:

> *"In a sense, the earth is mounting an immune response against the human species. It is beginning to react to the human parasite, the flooding infection of people, the dead spots of concrete all over the planet, the cancerous rot-outs in Europe, Japan, and the United States, thick with replicating primates, the colonies enlarging and spreading and threatening to shock the biosphere with mass extinction. The earth is attempting to rid itself of an infection by the human parasite."* [32]

We Have a Choice

But, and let's put that BUT in big letters, we have a choice. We can decide to look after our planet, and the good news is that every person can play some small role in this. We don't need a government leader to tell us to stop throwing away plastic garbage (with plastic water bottles being a major form of ocean pollution threatening the life of fish). Every person can take a responsible approach to recycling, to lowering the amount of energy used in the home (which would directly or indirectly save freshwater), to ensuring that our cars do not emit unacceptable amount of emissions. We can plant trees instead of pouring more concrete.

We can all attempt to educate ourselves on pollution-free renewable sources of energy such as solar energy (and then vote for politicians who attempt to support such endeavors). It's estimated that one hour of sunlight hitting our planet delivers more energy in that one hour than all humans consume worldwide over the course of a complete year.[33] While we can't harvest all this energy from the sun, this does provide us with a glimpse of the potential that solar energy holds for us (and solar energy is a clean renewable source of energy). It's very likely that humans in the future, if our species still exists, will look back on us as being very careless in our dependence on non-renewable polluting fossil fuels.

Interestingly while some countries such as the U.S. are currently attempting to protect the dated jobs of coal miners (with the burning of coal being one of the greatest polluters of harmful carbon dioxide into our atmosphere), a recent statement from the U.S. Bureau of Labor stated that "the greatest single fastest-growing job in in the U.S. is solar photovoltaic installer, and the second fastest growing job is wind-turbine technician."[34] Wouldn't it make sense to retrain coal miners to work in alternate energy source jobs?

What we really need is for voters to support politicians who accept climate change and support the implementation of solutions, and we need to understand that as individuals each of us can make a positive difference in the fight against climate change.

Saving Our Planet as a Source of Meaning in the 21st Century

Although global warming is a huge problem that threatens our very existence, the crusade against climate change can contribute to help people find meaning in the 21st century. The fight against global warming is one that every person on the planet can take part in whether as an individual or as part of a more formal group. Saving our planet is a worthy goal. It is a goal that anyone can help to achieve. When people fight for a cause (usually in a war), this bonds people together and gives meaning to waking up each day. The fight against global warming could be universal. It could bring humans across the planet together in a common purpose.

Preventing Global Warming

Let's end this chapter by taking a look at some thoughts on how to stop global warming and other environmental concerns that threaten the health of our planet (and our health as well). This is not a comprehensive list by any means, but I hope it will show each reader that we can all participate in this fight, and the outcome could be very beneficial to us and future generations, and provide a sense of purpose for all humans in the 21st century.

It's been proven that people are happier when they feel they are somehow making a positive difference in their community each day. What could be more meaningful than creating a change that will help to leave our world a better place for our children and in the process help to make our world a better place now.

Reducing Carbon Emissions

A great deal of harmful carbon dioxide that is released into our atmosphere is directly related to the burning of fossil fuels (such as gas) that help to create energy sources to run our cars, our furnaces, our air conditioners, our factories, and so on. The U.S. Department of Energy states that in 2012 the average cost of energy per person was $3,052 annually in the U.S. Finding ways to become more energy-efficient can save people money and also help to save our planet. In New York State, as an example, it's estimated that 29% of all the energy used in the state comes from residential housing.[35]

You might be unaware that the energy used in an average home creates twice the amount of pollution as the average car (although we need recognize that we have far more cars than homes, thus making cars a greater problem overall).[36] Turning your heat down 1 degree in your home in winter might be negligible in terms of your comfort, but if millions of people did this, this could lower our energy usage. Similarly, if air conditioners were set 1 degree higher in the summer, this could also have a similar impact. Installing a programable thermostat reduces energy consumption when you're not home or when you're sleeping. This can save energy usage, helping to prevent global warming, while saving you money in energy costs.

Weatherizing your windows and doors can lower energy consumption and over time pay for the work done in energy cost savings (and many countries and states offer rebates for people for energy-efficient home improvements and also free energy audits to help you identify what you can do to create a more energy efficient house).

Of course, some readers might be thinking the obvious—what if I am the only one who does this? Fortunately, there are already many people attempting to reduce the amount of energy used on a personal basis (if for no reason other than to save money). David Suzuki, Scientist and Environmentalist, states:

> *"In a world of more than 7 billion people, each of us is a drop in the bucket. But with enough drops, we can fill any bucket."*[37]

Light Bulbs

Replacing light bulbs in our homes and factories with LED lights reduces heat that emits from incandescent light bulbs (which waste 95% of the energy they consume) and can also reduce your energy costs (not to mention the cost savings over time because your LED lights will last significantly longer than incandescent lights). In the process you are reducing heat being transferred back into the environment. The cost savings and energy reduction for businesses using hundreds of lightbulbs could be very significant.

The U.S. Environmental Protection Agency reports that if every American home replaced just one lightbulb with an Energy Star-rated light bulb, this would reduce greenhouse gas emissions by 9 billion pounds, or about the equivalent of the pollution from 800,000 cars.[38]

Automobiles

A NASA report concluded that motor vehicles are the greatest net contributor to atmospheric warming now and in the near term.[39] It is estimated that 29% of all energy-related carbon dioxide in the U.S. comes from gasoline-powered vehicles. Carpooling to work or making better use of public transit or even riding a bike can all lower harmful gases that are emitted from cars. Fuel-efficient cars along with electric and hybrid cars can also make a difference. As well, people can replace their gas lawn mowers with electric or rechargeable ones.

Increasing Natural Vegetation

While many people are aware that increasing carbon emissions in our atmosphere are contributing to climate change, most people fail to realize that natural vegetation across our planet can reduce these carbon emissions by converting them to harmless oxygen. When we think about climate change, the problem isn't just that we are pouring carbon dioxide into the atmosphere, it is also related to stripping natural vegetation to expand our cities, to mine the land, or to use the lumber for other purposes. Drastic amounts of rainforests throughout the world have been destroyed over the past century. It is no coincidence that harmful carbon dioxide has increased in our atmosphere at a time when so many forests have been destroyed.

More and more as I walk through my neighborhood, I see people landscaping their properties with decorative stones or concrete. We need to be replacing concrete with grass wherever possible, and we also need to be planting more trees, solutions that many people can assist with.

I live in a province in Canada where the provincial government has recently voted to end a moratorium of developers building on our protected "green belts" throughout the province. Such thinking is incredibly short-sighted (and yes, we currently have a conservative government backed by conservative Christians).

Voting for the Right Politicians

Of course, on a larger scale, those of us who live in democratic countries can vote for politicians who are committed to decreasing dependence on fossil fuels and increasing dependence on renewable energy sources. And when we are unable to make an impact at a national level, then we can begin our efforts in our own communities. Even with a current American administration that is more concerned with the economy than climate change, there are states, cities and even towns across the country who are committed to reducing energy costs.

Former New York City mayor Michael Bloomberg, as an example, has offered $70 million dollars in technical assistance funding for major cities across the U.S. to develop plans for running their transportation systems and buildings cleaner and more efficiently.[40]

What Are Other Countries Doing?

Outside the U.S., several countries are moving towards a greater use of renewable energy sources. For example, in 2015, 53.9% of Sweden's gross energy consumption came from renewable energy sources. Denmark has set a goal of converting completely to using only electricity and heating that is generated by renewable sources by 2035. It's estimated that Denmark will be able to generate all its electricity needs from wind energy alone and will have enough left over to contribute to the needs of Germany and Sweden.

In Europe, as a whole from 2005 to 2015, the share of renewable energy sources in relationship to the total amount of energy used increased from 9% to almost 17%.[41]

Nuclear Energy

While nuclear energy is still a significant source of energy in some countries (26.5% of the energy created in the European Union), there are concerns about the safety of using nuclear energy, although some experts would argue that these concerns have been exaggerated. These concerns include the possibility of a terrorist group destroying a nuclear facility and causing enormous harm from the radiation fallout. Dangers also include the possibility of natural disasters destroying a nuclear power plant such as the Fukushima disaster. While there are ongoing advances in helping to make nuclear energy safer, public opinion may prevent nuclear energy from reaching its full potential.

On the other hand, nuclear fusion is a far greater possibility for future energy needs. Nuclear fusion does not create any toxic waste. It is much safer to use. The problem at this time is that scientists have not found a way to commercialize nuclear fusion. It could be strongly argued that if a number of leading industrialized countries cut back a little on their military expenditures and targeted this money for further research into nuclear fusion (and other renewable sources of energy), we could make a significant dent in stopping the rise of global warming. The major challenge here is finding politicians who are willing to put the environment before their own ambitions (and the day has already arrived in many parts of the world that for politicians to stay in power, they will have to recognize and act on the public's concerns about climate change).

From Sun God to Sun Savior

For much of human history our ancestors worshipped the sun. In the 21st century the sun may very well be the savior of humankind.

Let's take a brief look at solar energy which would be a great clean renewable resource if we could make better use of it. Solar energy is a form of energy that can be harnessed by individuals as well as being a form of energy that can be implemented on a much greater scale.

Driving through neighborhoods in industrialized countries, it is not unusual to see a growing number of solar panels on the roofs of houses (often to heat swimming pools), and in driving through rural areas you will often see banks of solar panels in a field near a house or farm. As the cost of solar energy falls, more and more individuals will look at the possibility of implementing a solar energy to meet their needs.

Companies like Elon Musk's SolarCity and SunTegra have developed solar roofs to meet household energy demands which shows that individuals can begin to make better use of solar energy. In some countries you can even go to a store like Ikea and purchase solar installation kits for your home. Obviously one of the biggest obstacles with solar energy is the energy is only produced during the day when the sun is shining, but with advances in storage batteries this problem may be eliminated in the future.

At the current time, China, India and the U.S. are the leading solar energy producing countries, although this can be misleading as 66% of China's energy demands comes from the burning of coal which is one of the most harmful fossil fuels. In fairness to China, it is the world production leader in using renewable energy (14% of their total energy consumption comes from renewable energy sources) and it has plans to significantly increase its use of renewable energy.

Worldwide the use of solar energy is increasing adding almost 30% more solar energy production in 2017.[42]

Atheists, Christians, China, and Pollution

Not only is China attempting to increase its use of clean energy, it is also committed to reducing pollution. Beijing, which has been infamous in the past for its poor air quality, has significantly reduced the pollutants entering its surroundings. China's current goal is to see its good air days in the Beijing area to reach over 80%. In addition, they are attempting to reduce sulfur dioxide and nitrogen emissions by 15%. Currently China has hired more than 18,000 inspectors who will visit 80 cities in the country to enforce government policies related to reducing pollution.[43]

It's interesting how a country like China that is often considered to be a godless nation because of their swelling ranks of atheists (and restrictions on religion) is concerned about the health of our planet while a country like the U.S. has a major political party (supported by its Christian evangelical base) that turns it back on regulations to reduce pollution and incentives to increase clean energy sources whenever it is in power.

Other Renewable Energy Sources

There are a number of other renewable energy sources being pursued in various countries. These include harnessing the tides and creating microbe-based biofuels. In addition, in fighting global warming it may become necessary to use geoengineering to alter our climate or reduce harmful gases in our atmosphere. For example, giant sun screens could be placed in our atmosphere to redirect some of the sun's rays away from our planet. It's even possible that scientists will discover the mechanics involved in plants absorbing carbon dioxide and converting this unhealthy gas to oxygen. Finding a way to imitate plants might help us to reduce the carbon dioxide in our atmosphere.

Although there are some countries and areas within countries that might lag behind others in doing their part to fight global warming, progress can still be made. At the time of writing this book the U.S. has both a president (Trump) and ruling party (the Republicans) who don't believe in climate change and who are actually eliminating regulations from past administrations that were in place to help fight climate change. In spite of this lack of leadership, some individual states within the U.S. are implementing their own plans to tackle climate change. For example, the state of California has set a goal (and officially voted for this to occur) that all electrical energy will come from clean-energy sources such as wind and solar by 2045.[44]

What Can We Do as Individuals to Stop Global Warming?
From an individual standpoint here are some other suggestions (in addition to what has already been explored) we can all do to reduce global warming:

- *plant a tree or a garden*
- *green your property with more lawn, gardens, and less concrete*
- *ensure that your car and home furnace are regularly tuned-up*
- *walk more, drive less*
- *turn off lights and appliances when they are not being used*
- *buy local food (the greater the distance your food has to travel to reach you, the greater the pollution that results from the delivery vehicles)*
- *reduce your water usage such as taking shorter showers (it takes significant energy to produce clean water)*
- *unplug devices that you don't use on a regular basis*
- *keep the tires properly inflated on your car; this helps to reduce the amount of gas that you use*
- *purchase only Energy Star rated appliances and devices*
- *recycle*
- *educate yourself on global warming and renewable energy sources*
- *speak up; let your politicians at every level know that global warming is a concern that needs to be addressed*

Natural Selection in the 21st Century
The general theme of this book is exploring our search for meaning in the 21st century. As humans, for much of our time on this planet we have faced problems of disease, war and hunger. For much of our time, our species has faced a brutal fight for survival, often against each other. The 21st century has ushered in an era of prosperity that humans have never experienced to such a degree

before. Yes, we still have minor wars. And yes, there is still poverty and disease. But the extent of these concerns is significantly less than at any other time in human history.

The 21st century, as explained in more detail in the first chapter of this book, has brought us more apparent comfort and security than we have ever experienced before. While it could be said that the 20th century initiated the beginning of positive changes for humanity, that was also a century punctuated by two major world wars and the Spanish Flu as one example of a disease epidemic.

The real question in the 21st century might become whether we can hold onto our new prosperity, bringing all people on our planet onboard, or whether we are going to ignore the warning signs around us and fall into a false security, a false security that will result in the destruction of all the advances we have made, and continue to make, in living standards throughout the world.

Climate change is perhaps the greatest challenge facing us in the 21st century. If we don't address this problem now it will create an even more serious concern for our children and grandchildren. The World Health Organization estimates that currently 12.5 million people die each year directly as a result of climate change and that number is growing significantly each year.[45]

Throughout this book we have looked at the power and influence of natural selection. The bottom line is that every creature on this earth is in a constant battle for survival. For every species of animal or plant on this planet, meaning is primarily rooted in successfully passing on its genes to ensure the continued survival of that species. Unlike every other species on our planet, we have a greater amount of control over our future.

To a very large degree we have a choice about our survival, but make no mistake about it, throughout our history we have made some horrible choices (think of wars as a prime example). A major question facing us at this juncture in our history is whether we can come together as a species to fight a concern such as climate change. It is our choice. It is our future. Left unchecked, climate change will become the great driver of natural selection in the future, and as a species we may very well follow in the footsteps of the dinosaurs unless we address this problem.

Tackling the problem of climate change can provide meaning

for individuals and collectively for much larger groups (even countries). This is a problem we can all help to fight, and for one of the few times in our history, this is a fight where we will be saving lives and contributing to a better standard of living throughout the world, instead of trying to exterminate each other.

A recent report from the Nobel Prize-winning Intergovernmental Panel on Climate Change states:

> *"Preventing an extra single degree of heat could make a life-or-death difference in the next few decades for multitudes of people and ecosystems on this fast-warming planet."*[46]

In a local newspaper this morning, I counted seven articles directly related to the problem of climate change. Awareness of this problem is increasing every day. The major question now is who is going to get onboard to take action? Can the fight against global warming unite humans in a common cause in the 21st century?

Climate Change Update

It's two weeks before the publication of this book and today more than 1.6 million students in more than 120 countries, representing 2,233 cities worldwide, walked out of their classes to protest adult inaction related to climate change.[47]

At the same time, even in the U.S., increasing attention is swinging towards addressing climate change with the proposal of a "Green New Deal."[48]

Change is coming.

- 10 -
Survival of the Fittest

A few days ago, a baby bird appeared on the lawn in our backyard. Although the bird had its eyes open, its wings were not fully formed so it couldn't fly. The best the baby bird could do was hop around although for the most part it stood still, looking upwards to the sky above as though it was praying, its mouth open awaiting its next meal from one of its parents, its incessant chirping calling for help, but more likely attracting a predator.

If the baby bird had been in a Disney movie, one of its parents would have swooped in to rescue it with heart-warming music reflecting our love for the beauty of nature. Or perhaps, in a Disney movie another animal might have rescued the baby bird. Maybe a cute bunny would have appeared to care for the hungry and fragile bird. And yes, I enjoy watching these movies despite their unrealistic portrayal of nature.

The appearance of the baby bird reminded me of a similar experience when I was a child. Finding a baby bird in our backyard, I asked my parents if I could bring it into the house to care for it. They insisted that we should leave the baby bird alone outside and let nature look after it. Reluctantly accepting their wisdom, I went to bed wondering what was going to happen to the baby bird.

The next morning, I rushed out of the house to see how the baby bird was doing. What I found was a scattered mess of feathers, part of a head, a claw, and a small amount of a bloody mess that was once the small bird. Yes, nature most certainly looked after the baby bird.

Nature doesn't care about cuteness or feelings, or the happy endings of Disney movies; it ruthlessly cares about the survival of the fittest. But within nature we find some instances where natural selection favors cooperation instead of destruction. And for us as humans, cooperation rather than aggression may hold the key to our future survival and provide a foundation for finding meaning in the 21st century.

Could Survival of the Fittest Depend on Cooperation Within a Species?

Any species that continues to thrive does so because it is the fittest or most adaptable. Unfortunately, the phrase survival of the fittest is often interpreted as survival of the physically strongest which is not necessarily true. As mentioned in previous chapters, for any species to survive it is necessary to pass on a genetic blueprint (DNA) that favors the survival of that species. The fact that the mightiest of dinosaurs no longer exist provides strong evidence that the survival of any species is not necessarily rooted in physical strength, but rather in adaptableness.

In the introduction of the revised version of Robert Axelrod's book *The Evolution of Cooperation*, Richard Dawkins writes:

> *"As Darwinians we start pessimistically by assuming deep selfishness at the level of natural selection, pitiless indifference to suffering, ruthless attention to individual success at the expense of others. And yet from such warped beginnings, something can come this is in effect, if not necessarily in intention, close to amicable brotherhood and sisterhood."*[1]

While the progressive march of natural selection over billions of years has often been unrelenting in determining which forms of life survived, there have also been pockets of survival which favored cooperation rather than a seek-and-destroy approach (and growing interest and research in this area is discovering that the impact of cooperation on the survival of the fittest might be much greater than initially thought).

Cooperation as a Survival Mechanism

A few minutes ago, I fed a pair of clownfish in one of my basement aquariums. Most people would recognize the orange clownfish, with their white stripes bordered by black, from the popular

SURVIVAL OF THE FITTEST / 219

movie *Finding Nemo*. What some people may not know is that clownfish have what is known as a symbiotic relationship with poisonous invertebrates called sea anemones.

When most fish, other than clownfish, brush against the tentacles of an anemone in the ocean, the tentacles fire a harpoon-like filament into the fish and inject a paralyzing neurotoxin. The tentacles then guide the paralyzed fish into the mouth of the anemone which is set in the center of its tentacles. Sort of like Audrey in *A Little Shop of Horrors*, only in the ocean. The waving, often beautiful, arms (known as polyps) of sea anemones lie in waiting for a fish to pass close enough to become trapped in their venom-filled tentacles.

At first glance, it would appear that poisonous sea anemones have survived for millions of years by having developed a system for murdering their victims. But hold on, there is more. Some forms of the sea anemone allow clownfish, of several varieties, to live within the poisonous tentacles without ever being harmed. Over time the anemone and clownfish have developed a form of cooperation known as symbiosis. Anemones provide protection to the clownfish from larger fish and they even clean harmful parasites off the clownfish. In return, clownfish bring scraps of food into the waiting tentacles of the anemones. Both creatures benefit from cooperating with each other.

In addition to cooperating with clownfish, sea anemones sometimes hitchhike on the backs of hermit crabs. The crabs provide leftover food for the anemones as well as protecting the anemones from predators such as starfish and fireworms. On the other side of this cooperative agreement the poisonous tentacles of the anemones protect the crabs from predators such as fish and even octopuses.

When a crab finds an anemone attached to a rock, it will poke its tentacles until it releases its grip on the rock, after which it will hold the anemone in place on its shell until the anemone attaches itself to the crab. The crab then wanders around the ocean floor with its own bodyguard.

In another example of cooperation in the ocean, snapping shrimp and goby fish have developed a cooperative relationship. As the almost blind snapping shrimp digs a tiny hole on the ocean floor to use as its home, the goby fish stands guard while the shrimp digs.

If a large fish wanders by, the fish wiggles its tail against the antennae of the shrimp warning it to hide. Once the home in the sand is completed, the shrimp allows the goby fish to hide in the hole or even sleep in the hole with the shrimp, thereby protecting the fish.

Cleaner fish, such as wrasses, provide "cleaning stations" for larger fish. These small, often narrow, fish can even be seen swimming in and out of the mouths of larger fish without ever being harmed. The cleaner fish benefit by eating parasites, mucous and dead tissues from the host fish, while the larger fish benefit from having potentially harmful parasites removed from their bodies.

Are We a Host Like a Sea Crab?

Within our own bodies, there are an estimated 100 trillium forms of bacteria and other microbes. The cells of these bacteria outnumber our own cells by about ten to one, and their genes outnumber our genes by about 100 to one. It could be argued that we are simply the host for trilliums of microbes in the same way that a crab might be a host for the sea anemone. And while some of the bacteria that live within us may be harmful to us, other forms of bacteria such as Lactobacillus (frequently found in yogurt), are helpful.

We could spend many more pages looking at how cooperation between various forms of life contributes to survival with natural selection in these cases favoring cooperation rather than physical strength and aggression, but hopefully I have provided enough examples to make the point that cooperation among different creatures can be a factor in the survival of some forms of life.

Altruism

Another word that will be introduced in the following discussion is altruism. Cooperation is often defined as working together towards a common goal (in the current discussion this common goal would be the survival of a species) while altruism might be defined as an unselfish regard for others (even to the point of sacrificing one's well-being to increase the chances of survival for someone or something else).

In nature most creatures don't care for their young. Instead most animals depend on producing huge numbers of offspring with the strongest surviving and the weak perishing (sometimes even being cannibalized by their siblings). Where animals do care for their young, it is often for a relatively brief period of time until their offspring have been taught how to hunt and avoid enemies. The exception to this would be as we move up the primate scale where orangutans, as an example, care for their young for up to eight years.

There are exceptions to this general rule. There are in fact some animals where parents are willing to sacrifice their lives to protect their offspring. For example, female caecilians (that look like worms but are really amphibians) allow their offspring to eat the skin off their bodies (which regenerates on the mother every three days). A black lace weaver spider mother offers herself to her children in a cannibalistic feeding in order for them to survive. If a pseudoscorpion mother fails to find food for her young, she will allow them to suck the nutrients from her body until she is dead.

Female killdeers (a small bird) will pretend to be hurt and lead any intruders away from their nest of eggs or babies, often exposing themselves to danger.

Within some specific species, animals will cooperate to protect related members. For example, many birds will signal a warning if a predator is nearby even though this warning may place the life of the signaler in danger. Some birds will even attack much larger birds or other animals who attempt to eat the eggs or young in their nests.

As elephants move, they protect their young by keeping them in the center of their herd, sometimes even in bodies of water when threatened by crocodiles. A mother giraffe will keep her calf in between her legs if lions attack and she will use her long legs to fight off the predators, risking her own life when she might otherwise use her long legs to flee.

Some animals care for their young to the point of offering their own lives to save their offspring (altruism), while other creatures abandon their young and let them fend for themselves. While we have been looking at some examples of how altruism and cooperation play a role in the survival of some animals, let's briefly be reminded that this is not the norm amongst most animals.

Far more animals simply abandon their young. For example, reptiles lay their eggs and then depart. Even when a snake gives live birth, she immediately leaves the baby snakes. Many forms of fish (including the cute clownfish) will protect their eggs, only to eat some of the tiny fish when the eggs hatch. Some animals, such as the loveable pandas, will often choose only one baby to care for, leaving the other (pandas generally have twins) to perish (sometimes even killing it), much to the horror of the public when this occurs in a local zoo. Although the examples in this paragraph directly relate to the survival of the fittest, they are not an example of cooperation or altruism.

Does the Survival of Our Species Best Depend on Physical Domination or Cooperation?

We might ask where we fit into this. While we tend to care for our young—and there are countless stories of parents sacrificing their lives to save their children, and records of complete strangers sacrificing their lives to save others—we are also a species that has throughout history engaged in wars and mass genocide, and infanticide (and sometimes even cannibalism).

If we are to survive as a species (and as I suggested before in the previous chapter, our planet doesn't care whether we thrive or whether we become extinct), we might ask how natural selection is going to play out in our future? Will our species survive if we continue to engage in wars or ignore global warming, or will we have better odds of surviving as a species if we turn instead to cooperation? In the end, it will be natural selection that determines our fate, but for humans will natural selection favor cooperation or aggression? Unlike other forms of life on our planet, we can consciously decide our answer to this question.

Homo sapiens (that's us) emerged approximately 300,000 years ago, and each year this date has been moving back further in time as archaeologists make new discoveries (and our most direct relatives can be dated back several million years ago). There were at least seven species of the Homo genus (our relatives), any of whom might have become humankind as we are known today. Given that Homo sapiens do not appear (from the fossil records) to have the physical strength of some of our cousins such as the Neanderthals, how is it that we survived, and they perished? Is it possible that among our various relatives that Homo sapiens were the one group that made the best use of cooperation? And while this cooperation might have aided our survival though increased effectiveness in hunting (and eventually led to the agricultural revolution), it undoubtedly also contributed to being more effective in slaughtering our closest relatives.

Currently we are faced with a dilemma where a major war or problem such as global warming could seriously deplete our species. Cooperation in preventing wars and dealing with climate change could improve the standard of life worldwide for humans and contribute in a very positive manner to our long-term survival.

John Hands, in his book *Cosmo Sapiens—Human Evolution from the Origin of the Universe*, writes:

> *"Collaboration is a more significant cause than competition in the development and survival of organisms and is extensive at every level of life."*[2]

Religion and Violence

History has shown that religion is not the answer, at least in its current form, to create an atmosphere of worldwide cooperation. In fact, religious differences may have contributed more to suffering, persecution, and wars than any other single cause. Recently I read an article stating that the two world wars of the 20th century that inflicted the slaughter of millions of lives had nothing to do with religion. Perhaps the author of such thinking should visit a holocaust center before making such a statement. In an interview, the noted religious historian Philip Jenkins stated (in reference to World War 1):

> *"Throughout, and in every country, the war was presented as a holy war, a cosmic struggle. The war was fought by the world's leading Christian nations, and on all sides, clergy and Christian leaders offered a steady stream of patriotic and militaristic rhetoric. Many spoke the language of holy war and crusade, of apocalypse and Armageddon."*[3]

For many years, religion has generally been the cause of discord, rather than a foundation for peace. When one faith sees itself as being superior to that of another (which is the case of every major religion), this results in believers attempting to convert the unconverted to the "true" way (and throughout history, warfare and persecution have been seen as a legitimate means to convert non-believers). Unless a faith arises in the future that is more universally accepting and accepted, and is based on cooperation rather than an attempt to dominate, religion will not provide the answer to prevent us from extermination.

Although beliefs in God and gods have constantly evolved throughout history, it's unlikely that an older faith will evolve fast enough (and be accepted by enough people) to save us from our current potential species-ending problems.

While the faithful believers of whatever God their religion subscribes to would like us to believe that cooperation (often voiced as "love") and morality are the by-products of their religion, this is not necessarily true. In his book *Undeniable— Evolution and the Science of Creation*, Bill Nye writes:

> *"Altruism is not a moral or religious idea, no matter what some people might tell you. It is an essential, biological part of who or what we are as a species."*[4]

Does Religion Make Us More Moral?

Some would argue that religion gives us morality, that religion is the foundation for reaching out and being kind to others. A study, widely reported, and quoted here from the *Huffington Post*, found:

> *"There were no major differences between Christians and those who do not believe in God in their moral attitudes towards compassion for those who are suffering and ensuring justice and fairness for all."*[5]

It would even appear that morality can be found among some other animals. If this is the case, then morality has nothing to do with religion. If any non-humans are found to have a sense of morality, and non-humans do not have any Holy Scriptures to provide direction for them, then it cannot be argued that morality is rooted in religion. Frans De Waal, named by *Time Magazine* as one of the world's most influential people (and a leading expert on primate behavior), states:

> *"Female chimpanzees have been seen to drag reluctant males toward each other to make up after a fight, while removing weapons from their hands. Moreover, high-ranking males regularly act as impartial arbiters to settle disputes in the community. I take these hints of community concern as a sign that the building blocks of morality are older than humanity, and that we don't need God to explain how we got to where we are today."*[6]

Taking the previous conversation a little further, we find that within the United States (and this trend has been confirmed in other countries) that the states with the highest murder rates tend to be the most highly religious states, and the states with the lowest murder rates tend to be among the least religious in the country.

Another interesting fact is that atheists are significantly underrepresented in the U.S. prison population and have a lower divorce rate than religious Americans. A Canadian study found that conservative Christian women had higher rates of domestic violence than non-affiliated women. It has also been found that the most secular (non-religious) nations in the world consistently report the highest levels of happiness along with the highest level of charitable aid.[7]

Various religious faiths preach that our laws and our sense of morality have come directly from God. In reality, our laws have evolved over time with courts and governments both defining and interpreting them. The laws that are a part of any country today have slowly emerged in incremental steps. While religions may have contributed to our laws, they are not the sole source, and it could be argued that religion often has a negative effect on our laws. In the *Huffington Post*, Camille Veselka writes:

> *"Nowadays few countries are firmly based on religious beliefs, yet in many countries religions affect certain aspects of the law. Allowing religions to affect laws limits people's freedoms.*

> *Religious influence should remain out of laws and*
> *out of politics in general."*[8]

While I'm not suggesting that atheism is the solution for world peace (or the saving of humanity), I am arguing that religion, at least in its current form, is not the answer either. As Steven Pinker writes in his bestselling book, *"The Better Angels of Our Nature"*:

> *"We sympathize with, trust, and feel grateful to those*
> *who are likely to cooperate with us, rewarding*
> *them with our own cooperation."*[9]

The basis of cooperation is not found in religion. As outlined previously in this chapter, there are animals that exhibit various forms of cooperation to survive, and they certainly don't belong to any religious faith. Cooperation for these animals has been established in their genes as a way for survival. Similarly, cooperation as a survival mechanism for the human species may be hardwired into our genes. By ignoring our need to cooperate, we may hasten our demise; by heeding our need to cooperate, we may enhance our survival.

The Power of Cooperation in the Survival of the Fittest

At this juncture in history, it could be argued that cooperation among humans would not only improve the standard of living across the world, it could also save humanity from extinction (from nuclear or biological wars, or climate change). And the good news is that cooperation can begin on a small scale and spread. As Robert Axelrod writes in his seminal book *The Evolution of Cooperation*:

> *"...cooperation can get started by even a small cluster*
> *of individuals who are prepared to reciprocate*
> *cooperation, even in a world where no one else*
> *will cooperate. The analyses shows that the two key*
> *requisites for cooperation to thrive are that the*
> *cooperation be based on reciprocity, and that the*
> *shadow of the future is important enough to make*
> *this reciprocity stable. But once cooperation based*

SURVIVAL OF THE FITTEST / 227

on reciprocity is established in a population, it can protect itself from invasion by uncooperative strategies."[10]

In looking at our evolutionary history, cooperation may have played a greater role in natural selection than many people tend to think. Martin A. Nowak, from the Program for Evolutionary Dynamics at Harvard University, writes:

"Cooperation is needed for evolution to construct new levels of organization. The emergence of genomes, cells, multi-cellular organisms, social insects and human society are all based on cooperation. Cooperation means that selfish replicators forgo some of their reproductive potential to help one another."[11]

During the past century, for the most part evolutionists emphasized that natural selection was rooted at an individual level—the fittest individuals survived, passing on their genes to their offspring, resulting in the strengthening of their species. In the latter part of the 20th century it became more common for evolutionists to accept that cooperation may have played a role in the future of any species. From an evolutionary perspective, there are social behaviors among various animals that contribute to the future survival of that species. Along the same lines, when humans help other humans, this would likely be beneficial to the future of our species.

In his book *The Descent of Man*, Charles Darwin wrote:

"A tribe including many members who, from possessing in high degree the spirit of patriotism, fidelity, obedience, courage and sympathy, were always ready to aid one another, and to sacrifice themselves for the common good, would be victorious over most other tribes; and this would be natural selection."[12]

Live and Let Live

Earlier in this book it was outlined how during both the 1st and 2nd World Wars, there were examples of spontaneous ceasefires along combatant lines on Christmas day. Actually, there were frequent instances where troops agreed to stop fighting throughout the year and not just on Christmas day.

Tony Ashworth, a British sociologist, using diaries and letters found widespread examples of joint ceasefires by soldiers in the trenches during the 1st World War that were determined by the soldiers instead of superior officers. These soldiers often demonstrated mutual restraint in a form of cooperation, often during meal times when everyone stopped shooting. Soldiers were even known to stop fighting when the weather was bad, and when the weather improved there are many instances where the ceasefire simply continued. At other times, one side established routines for shooting which allowed the other side to hide so they wouldn't be shot. One soldier noted:

> *"On a fixed hour each day the Germans blew the same portions of the line to dust with minenwerfers, our men having departed half an hour previously, according to the established routine from which neither side ever diverged."[13]*

This created a culture that has often been described as "live and let live." On a very basic level, it was more difficult to shoot someone on the other side once some form of personal contact had been established.

Such cooperation was terminated when army leaders ordered raids on the enemy and required a monitoring of the results. The soldiers in the trenches often understood the futility of war. Unfortunately, the leaders of countries were often motivated by a zero-sum game where someone had to win and someone had to lose, even if millions combatants could end up on the winning side, but still lose their lives or be wounded physically or psychologically impaired.

The Power of Social Media in Contributing to Cooperation

In the 21st century social media could become the ultimate weapon against war although it could also create further problems with cyberbullying, cyberattacks, mobilizing mobs capable of perse-

SURVIVAL OF THE FITTEST / 229

cuting identified minorities, or recruiting new members for terrorist organizations.

At the time of writing this book, it's estimated that there are over 2.44 billion people across the globe who use various forms of social media. While protests throughout history have altered the actions of governments and sometimes even overthrown a government, there has never been a form of communication that is so instant and worldwide as social media being spawned by the Internet. It's estimated that every minute there are more than 30 million messages sent on Facebook and almost 350,000 tweets.[14]

British MP Douglas Carswell states:

> *"Everything that the Internet touches, it transforms. The barriers to entry come crashing down. The big government model that threatens to bankrupt and bully us is not just unaffordable; it is also increasingly impractical. The digital revolution is a coup d'état against the tyranny of this elite. It overthrows these second-hand dealers in other people's ideas."*[15]

Social media allows the average citizen to connect with any level of politician in a manner that has never happened before in history, or other similar minded individuals. It also provides an instant way to communicate news throughout the world. In a sense every person has become a journalist or politician. Even in countries where there is Internet censorship, people still find ways to let the outside world know what is happening and find ways to be informed about life (particularly living standards) in other parts of the world.

As Zaynep Tufekci, Faculty Associate at the Harvard Berkman Klein Center for Internet and Society, writes in her book *Twitter and Tear Gas—The Power and Fragility of Networked Protest*:

> *"The Internet similarly allows networked movements to grow dramatically and rapidly, but without prior building of formal or informal organizational and other collective capacities that could prepare them for the inevitable challenges they will face and give them the ability to respond to what comes next.*

By deploying technologies to effectively mobilize, movements can avoid many of the dreary aspects of political organization. There is real power here."[16]

Information shared across social media could help stop wartime atrocities and/or human rights violations as perpetrators of these crimes are identified, photographed, and videoed, providing proof of their transactions. Social media is helping to make people more accountable for their actions. It seems that almost every week there are new pictures posted online of police brutality, or violence against minority groups, or human rights abuses. These powerful images are becoming an effective way to inspire positive change.

On a broader scale, the demonstrations and protests known as the Arab Spring beginning in late 2010 and continuing throughout 2011, were often orchestrated by people connecting on social media, and while it has been shown that there were other factors involved, social media played a prominent part in these uprisings. In the end these protests led to regime changes in Tunisia, Egypt and Libya.

Marshall McLuhan, a Canadian university professor, coined the word "Global Village" in the early 1960s. In the 21st century the Internet has truly created a global village, a worldwide connective network that could be the vehicle for promoting greater cooperation. Whereas governments and religions have historically failed to promote peace across the globe, perhaps people can begin to reach across borders, embracing others and accepting their differences in the same manner that soldiers in the wars of the past sometimes put aside their weapons and questioned what they were really doing in killing other humans. It doesn't require a government or a religion to bring us together to save our planet and our species.

Cooperation as a Source of Meaning in the 21st Century

A theme that we have been exploring throughout this book is how 21st learning is impacting our search for meaning. From our most distant ancestors to our neighbors in the 21st century, cooperation has provided both survival value and meaning with the exception that cooperation in the past was often narrowly defined by the group (class, religion, race, ethnicity, and other factors) that one belonged to.

SURVIVAL OF THE FITTEST / 231

As we search for meaning in the 21st century, we can find meaning in altruism and cooperation, but we need to look beyond the borders, the walls, and the barriers we have placed between us and others (whether literally or figuratively) and expand our sense of cooperation to include all humanity.

Unfortunately, technology of the 21st century allows the military to destroy a faceless enemy from thousands of miles away. Drones, manipulated as though they were part of a video game, routinely attack targets throughout the world. Robotic soldiers may form armies of the future. While technology has the potential to save humanity, it also has the potential to destroy us.

For the first time since our ancestors began to walk in an upright position, we have the potential to consciously control our fate as a species. Individuals and groups can choose to aggressively enforce their false superiority (which often has its roots in a variety of religious faiths or political ideologies, each tending to claim that their way is the best way), or we can begin to move towards an acceptance of equal rights for all humans, and although I'm not going to be talking about this at length, we also need to consider the rights of other creatures who share this planet with us.

Progress

Recently in reading a book on 18th century England, I found the following passage:

> *"In eighteenth-century England the severest penalty was reserved for treason. The offender would be 'drawn to the place of execution on a hurdle', and there hanged, cut down while still alive, disemboweled, castrated, beheaded, finally quartered. Compared to contemporary French punishments, which reveled in red-hot pincers and drops of burning oil, this was relatively humane.*[17]

In 18th century England, prostitutes and homosexuals were often whipped or placed in stocks where they were at the mercy of a ruthless and unpoliced crowd who often stoned them until they died. Even children under the age of 10 could be hung for minor crimes such as the theft of a handkerchief. In most parts of the world we have come a long way since the 18th century (and the eras before this).

Although some politicians like to stir up their supporters with claims of increasing violence and the threat of terrorism, the reality in the 21st century is that far more people die worldwide from automobile accidents (30,296 deaths from car accidents in the U.S. in 2010 which is approximately 10 times the number of deaths in the World Trade Towers terrorist attack) than from wars or terrorist attacks, yet these same politicians keep pumping unbelievable amounts of money into the military while ignoring education and health care.

While humans have made incredible progress in reducing violence worldwide, we need to place less emphasis on military power and more emphasis on reaching out empathetic hands across borders.

Equal Rights for All Humans

In the 21st century can we ensure equal rights for all humans? Rights that would improve the quality of life for all humans and rights that would protect our future.

Since the earliest beginnings of life on our planet, cooperation has played a role, even if at first glance many lifeforms appeared to have survived simply by being the strongest.

If we are to survive, we must find a way to cooperate in spite of some out-of-touch political or religious leaders. In a recent book titled *War on Peace,* by Ronan Farrow, on the relentless assault on peace (especially by the current U.S. Trump administration, but traced back to previous administrations as well, and certainly not limited to U.S. administrations), we read:

> *"For the first time, an administration proposed removing 'just' and 'democratic' from the list of qualities the United States sought to encourage around the world.*[18]

We Can Drive the Change

For us to survive and prosper, change must come from individ-

SURVIVAL OF THE FITTEST / 233

uals. Grassroot movements need to impact government policies. We can longer assume that government or religious leaders know what is best for us. While it is true that natural selection will eventually favor the governments or religions that are most adaptable, we can no longer wait hundreds or thousands of years for this to occur. The time has come to put aside our differences and accept our neighbors across the planet, regardless of religious or political beliefs, regardless of skin color or ethnic background, regardless of culture, regardless of education, regardless of social or economic status, and regardless of age.

Through cooperation we can change our planet. We have the means to eradicate both disease and poverty. We have the means to feed every person. In the U.S., the military budget is eight times the budget for education ($595.5 billion for the military and $70 billion for education in 2016). What would happen if these numbers were reversed? Worldwide, in 2016, $1.686 trillion dollars was spent on arms. In 2013 the U.S. accounted for 39% of the total world spending related to the military (and the next ten countries combined only accounting for 21.2% with China having 9.5%, Russia 5.2%, and Japan with 3.4%).[19] What would happen if this money was spent on education or eradicating poverty?

What would happen if the U.S., Russia, and China demonstrated responsible leadership that emphasized cooperation throughout the world? What would happen if religions across the planet promoted cooperation instead of divisiveness?

In democratic countries, every individual has the power to create change. When voters push toward solutions for global or local problems, politicians will listen. In the end, the overwhelming majority of politicians support the laws and bills that will continue to get them elected. When the masses decide to only vote for politicians who will support solutions for issues such as global warming or poverty, we will see change happening, and we will be the drivers of this change.

When church leaders begin to understand that more and more people want beliefs that are accepting of other faiths, beliefs that are based on loving one another instead of being rooted in ancient superstitions, and beliefs that are based on personal responsibility for today instead of some pie-in-the-sky rewards in some distant tomorrow, then religion could be a part of meaningful change (and a positive part of contributing to meaning in the 21st century). When religious leaders see membership dwindling, which results in less money coming in, doctrines and ideology will change. Wouldn't it be nice if some religious leaders pushed for change before this occurred?

Do We Have the Desire to Survive?

We have the means to change our world and protect the future of our species. But to do this, we must have the desire. We need to passionately want to put an end to wars and inequalities. We need to passionately want to halt the carbon emissions that are ruthlessly contributing to global warming or the deforestation that is negatively impacting all life on our planet. We must monitor genetic engineering, artificial intelligence (AI), and future technologies that haven't appeared yet with an underlying concern of what is best for all humans, instead of what might profit a select few.

Cooperation can save humanity. Cooperation can create positive change across our planet which could see an end to inequalities and human rights abuses. Through cooperation, people can survive and prosper regardless of religious, political and cultural differences. And as we attempt to find meaning in the 21st century, whether on an individual or collective basis, cooperation can provide an answer for us to build on.

Random Acts of Kindness

We need more random acts of kindness and less reactionary words such as "Our thoughts and prayers are with the families of the victims." Over and over again, these thoughts and prayers make no difference in preventing future tragedies. While there may be times when the only words we can find to express our sorrow might be thoughts and prayers, wherever possible we need to turn these thoughts and prayers into action.

Throughout this book the power of natural selection has been explored. As previously mentioned, any lifeform is in a constant struggle for survival. In the end any species that continues to thrive does so because it is the fittest or most adaptable.

Although the phrase "survival of the fittest" is often interpreted as survival of the physically strongest, this is not necessarily true as had been explored in this chapter.

Soccer Team Trapped in a Cave

During the past few weeks as I have been writing this chapter, a drama in a remote mountainous area between Thailand and Myanmar has caught the attention of the world. Twelve boys, from a soccer team, and their coach had been exploring a local cave system known as Tham Luang, or the Great Cave of the Sleeping Lady, when monsoon rains flooded areas of the cave, trapping them deep inside the caves. Ten incredibly anxious days would pass before a pair of British divers made contact with the boys.

Every evening Buddhist monks visited outside the caves where they led the parents in prayer. Shamans from local tribes offered their prayers and even offered animal sacrifices to their gods. There is little doubt that people of a wide range of religions around the world offered their prayers for the successful rescue of the boys, a rescue that would prove to be very dangerous.

In the end it would be a team of professional divers who would lead the boys to their safety. As an experienced scuba diver, I have dived in underwater caves in Mexico, with careful preparation and years of experience. I couldn't imagine 12 boys and their coach, some of whom couldn't even swim, attempting to travel through narrow passages along the 2.5 miles it would take for them to reach the cave entrance from where they were trapped.

While I can see training someone for a few minutes of diving, the route that these kids and their coach faced would be a 6 to 8-hour journey (not all of it underwater). Along the way, they would have to be led through dark passageways sometimes filled with cold turbulent water where it would be impossible for them communicate their fears. Imagine the sense of claustrophobia that they would experience.

With cooperation from a range of countries, 13 International experienced divers and 5 Thai SEALS meticulously planned the escape, rehearsing what they were going to do and instructing each boy on what was going to happen. Although there is little doubt that Ekkapol Chantawong, the boy's coach who was a monk trained in meditation, helped to calm the boys during the ordeal, it was food and medicine that was brought through the cave system that kept them alive. Before embarking on the journey out of the cave, the boys were sedated to block their fears and the panic which might sabotage their rescue.

Sixteen days after the boys disappeared, the actual rescue began. Over a three-day period, the boys in groups of four, were tethered to professional divers. With two buddies for each boy, their journey began through some crevices that were so narrow that the divers had to take their equipment off to squeeze through. Three days later, the 12 boys and their coach were all safe.

The rescue was often described in the press as miraculous, yet in the end it was the expertise of the divers, not the hand of God, who saved the boys. If you had been one of the people trapped in the cave or one of the boys had been your child, would you rather have had the prayers of well-wishers during their 16 days horrific experience, or would you have rather had the expertise of the divers and the medical staff who assisted them?

The successful rescue of the soccer team is a great example of how people can transcend political or religious beliefs to work together to make a positive difference in our world.

Cooperative Action

If we are serious about ending problems such as persecution, poverty, hunger, the absence of education, and so on, prayers and thoughts are not going to provide the solution. Cooperative action, perhaps better spelled as COOPERATIVE ACTION will provide the solution if we allow ourselves to see all humanity as part of one big family. The expertise exists to solve many of our world problems, but unless we extend our hands out to each other, this expertise might as well be sitting at the mouth of a cave, unable to reach those trapped in danger. It will take action, not prayers and thoughts, to change our world.

Even the smallest acts of cooperation can lead to change. We don't have to sit back and wait for change from our leaders—in fact, based on history, that might be the worst thing we could do. Change will come from within. Change will come from one person connecting with another person, whether that person is part of your immediate family, neighborhood, workplace, or even online with someone you might never meet in person. Social media can allow us to connect with others throughout the world. When we put names and faces to the problems, we will begin to solve the problems.

Neither governments nor religions will likely provide the answers for humankind to survive. Each of us as individuals holds the key to the future of our species. Each person can find meaning in the 21st century by reaching out to help others, by reaching out to connect, by reaching out to work together to find solutions to the problems that we face. We are all in this together. The sooner we recognize this, the sooner we can move towards creating a world where every person enjoys the most fundamental rights—rights to adequate food, shelter, clean water and air, education, freedom of speech, equality, and personal security.

Unifying Through a Mutual Purpose

Throughout history, religious and political leaders have often unified their membership by declaring war on others simply because they have different beliefs or different cultural backgrounds. It has been said that there is nothing like a war to eliminate political differences that exist on either side. The major problem with such an approach is that wars are fought against other humans. Religious and political ideological differences have constantly resulted in the killing of our own species across the planet. For a creature that has supposedly evolved to the highest intellectual level of all animals on the planet, at times we have most certainly failed to use our intelligence in appropriate ways.

The Wars of the 21st Century

The wars of the 21st century must be on providing equal rights for everyone and on slowing the relentless pace of global warming. The wars of the 21st century can no longer be a zero-sum experience where we have winners and losers, where we have the victors and the vanquished, and where both sides suffer significant costs to their members. The wars of the 21st century need to be win-win where everyone involved benefits from a better standard of living.

Cooperation can help to provide meaning for both individuals and groups in the 21st century. Throughout history, people were often told that meaning came from worshipping or serving a god.

What would happen if this thinking was replaced with finding meaning in cooperation, in praising the best we can find in each other. As we will look at in the next chapter, this does not mean that humans have to abandon their religions, but it does mean that religions need to change their focus from the afterlife to the "now" life, and most definitely have to change their thinking of exclusivity to equality.

Perhaps we already have a predisposition towards cooperation through natural selection. As Matt Ridley writes in his book *The Origin of Virtue, Human Instincts and the Evolution of Cooperation*:

> *"Now an entirely new explanation for human society drifts into view. Maybe cooperation is such a feature of our society not because of close kinship, not because of reciprocity, not because of moral teachings, but because of 'group selection': cooperative groups thrive and selfish ones do not, so cooperative societies have survived at the expense of others. Natural selection has taken place not at the level of the individual but at the level of the band or tribe."*[20]

On a recent showing of *America's Got Talent* on U.S. TV, a 145-member choir named *The Angel City Chorale* from West Los Angeles, competed. Before the choir sang, their leader Sue Fink said these words:

> *"I wanted to start something to give back to the community. We tried to represent diversity, different religions, different ages, black, white, rich, poor, gay, straight, even the Republicans and Democrats can sit next to each other in our group."*[21]

This is a practical example of finding meaning and purpose in the 21st century. Incidentally the mission statement for this choir is "Building Community One Song at a Time."

Neil Turok, one of the world's leading theoretical physicists, writes in his book *The Universe Within*:

SURVIVAL OF THE FITTEST / 239

"Our global population has grown to seven billion and continues to rise. We are eating away at our supplies of energy, water, fertile land, minerals. We are spoiling our environment and driving species extinct every day. We are caught up in financial and political crises entirely of our own creation. Sometimes it feels as if the technological progress on which we have built our lives and our societies is just leading us towards disaster. There is an overwhelming sense that we are running out of time."[22]

Can we come together as a species to cooperate in finding answers to the serious problems that we face?

- 11 -
Why Are We Here?

The search for happiness and meaning has existed since the beginning of our development of consciousness (perhaps hundreds of thousands of years ago, and maybe even more). There are some who would argue that the search for meaning (and the resulting answers that are often found in religion) is hard-wired into our species. There might have even been some survival value for Homo sapiens ascribing meaning to supernatural forces.

Dr. Sam Harris, CEO and cofounder of Project Reason, in his book *Waking Up*, writes:

> *"The problem of finding happiness in this world arrives with our first breath—and our needs and desires seem to multiply by the hour. Seeking, finding, maintaining, and safeguarding our well-being is the great project that we are all devoted, whether or not we choose to think in these terms."*[1]

Roasting Marshmallows Around a Fire 100,000 Years Ago

A hundred thousand years ago, or so, as early humans sat around their evening fires, looking into the dark sky with endless twinkling stars, they must have wondered why they existed. Such a question was especially profound to their experience here on earth when life was so short and ruthless. Child birth often led to death for women and their babies. A toothache or an infected cut could lead to death. A wild animal, another human predator, lack of food, extreme weather, and the commonest diseases that we may easily ward off today could all contribute to an early death. For much of human history, our ancestors rarely lived much beyond their twenties.

With both suffering and death an everyday part of the lives of our ancestors, it's no wonder they invented gods to offer some hope to the daily hardships they experienced. Over time, the gods were given names such as Osiris, Ra, Zeus, Apollo, Mercury, and hundreds more. Eventually the many gods would evolve into one God, or would they?

One God, or are there Three?

Several hundred years after the death of Christ (when the Christian and Jewish religions had helped cement the belief that there is only one God), the Catholic Church (convened by Constantine in the year 325) would decide that God was really comprised of 3 gods: the Father, the Son, and the Holy Ghost (although this didn't become official church doctrine until the Council of Constantinople in 381 CE).

Actually, the belief in a trinity of gods long predated Christianity. The ancient Sumerians and the Babylonians both believed in the doctrine of a trinity (three persons in one god). In the Puranas, one of the oldest Hindu Bibles (written more than 3,000 years ago), we read, "O ye three Lords! The three gods, Brahma, Vishnu, and Shiva. Learn, O devotee that there is no real distinction between us."[2]

The ancient Egyptians also believed that there were three gods who exhibited power as one: Amun, Re and Ptah. And these are just a few examples of many ancient societies who believed that among their gods, there was a trinity of gods who joined together to form a supreme power.

So, is there one God, or are there three?

As explained earlier in this book, with thousands of different religions, many with their own interpretation of God, and with some people praying to Mary or the Saints as though they were gods, have we evolved back to the concept of many gods? Is there really one true God, or has each human invented an image of God or even gods that best meets their needs?

Jesus and the Fake Messiahs

Let's look at the beginning of Christianity (the world's number one religion with an estimated 2.2 billion followers currently). Christianity is based on the life of Jesus who lived around 2,000 years ago

WHY ARE WE HERE? / 243

(and one might note that there is no actual historical evidence that Jesus with his reported preaching and miracles even existed, other than what we find in the Bible). During the apparent time of Christ, we would find that this was a period when there were a multitude of gods and even dozens of hopeful messiahs who roamed the countryside claiming to heal the sick.

In her book *And Man Created God*, Selina O'Grady, writes:

> *"Nor could all religions adapt to the changing conditions created by trade. Cities killed off the old pagan gods. It is not for nothing that the early Christians called non-Christians pagani: the Latin paganus means 'country-dweller'. The pagan gods offered insurance against disaster, as long as they received an appropriate bribe in the form of a sacrifice. But they did not provide what the mystery cults such as Isis and certain other religions were beginning to offer: a way to live one's life, the promise of an afterlife and a personal relationship with god. Such comforts as these, which today are seen as defining elements of what constitutes a religion, were increasingly necessary to the alien world of the cities in the first century BCE."*[3]

Isis was a religion that had its beginnings with the ancient Egyptians. During the Roman era Isis grew in popularity. The central belief in Isis was the death and rebirth of Osiris which led followers to believe that they would be reborn in death and have eternal life (beginning to sound familiar?). In addition, followers of Isis believed in purification (with the priests bathing a person, asking their gods for forgiveness, and having water sprinkled on them, very similar to what we would find in many Christian churches today). Members of Isis believed in a personal relationship with their gods (similar to what we would find in Christian teachings today).

There were temples and priests devoted to public teachings about Isis. For a few hundred years Isis and Christianity competed against each other with Christianity eventually prevailing, although there is little doubt that Isis had a strong influence on Christian beliefs. Isis began to die out almost 400 years after Christ as various emperors began to give their support to Christianity.

During the 1st century CE, a Galilean became famous for his compassion and his miraculous powers. This man preached to the poor along the shores of the Sea of Galilee. Accompanied by his followers he wandered from town to town preaching and performing miracles. He created bread out of nothing and turned vinegar into oil. He spoke to his father in heaven. If you are thinking that I am describing Jesus, you would be wrong. I am describing Hanina ben Dosa, one of the more famous itinerant mystics who lived at the time of Christ.

Another contemporary of Jesus was Apollonius who some historians argue was even more famous at the time than Jesus. Apollonius's biography was written by Philostratus. Apollonius was reported to have supernatural powers, powers that enabled him to perform miracles, an apparent prerequisite for mystics who wandered the countryside during this period of time in history.

Biblical scholar Bart D. Ehman, writes:

> "Even before he was born, it was known that he would be someone special. A supernatural being informed his mother the child she would conceive would not be a mortal but would be divine. He was born miraculously. As an adult he left home and went on an itinerant preaching ministry. He gathered a number of disciples around him. He healed the sick, casted out demons, and raised the dead. His enemies delivered him over to the Romans authorities for judgment. After he left this world, he returned to meet his followers in order to convince them that he was not really dead but lived on in the heavenly realm."[4]

No, Ehman is not writing about Jesus; he is describing the life of Apollonius.

At the time of Christ, there were a number of other messiahs who wandered the countryside like him, performing miracles and preaching to the poor. Is it possible that without the support of a passionate leader such as the Apostle Paul (Saul of Tarsus) or the timely endorsement of Christianity by a ruler such as Constantine,

WHY ARE WE HERE? / 245

or the brutal imposition of their teachings by a powerful political organization such as the Catholic Church that any of these other messiahs could have become the foundation for a religion such as Christianity?

Throughout history there have been hundreds (perhaps even thousands) of people who thought of themselves as messengers of the gods or God. Some of these messengers founded major religions such as Christianity and Islam. In addition to major religions like Christianity and Islam, there are literally thousands of other religions, some that would fall under the umbrella of Christianity or Islam, and some that would be uniquely different.

Meaning from Worshipping God

For those who believe in gods or God (even if they have very little awareness of how their religion began), life's meaning generally comes from serving and worshipping him (or them). This is a core message of thousands of religions that have either existed in the past or present.

Of course, one might ask, why did God wait billions of years for humans to evolve on our planet instead of just creating us from the very start if he needed someone to worship him? Or, one might ask how could such a powerful all-understanding God be so needy that he had to create billions of people with the sole purpose of worshipping him? But for now, let's put such questions aside, although I suspect that in the 21st century, more questions along these lines are going to be asked.

The Rise of Unbelief

The main problem with traditional religions is that less and less people are believing their message. As Deepak Chopra writes in his book *The Future of God*,

> *"Faith is in trouble. For thousands of years religion has asked us to accept on faith a loving God who knows everything and possesses all power. As a result, history has walked a long and sometimes tumultuous road. There have been moments of great elation interspersed with unspeakable horrors in the name of religion. But*

> *today, in the West at least, the age of faith has drastically waned. For most people, religion is simply taken for granted. There is no living connection with God. Meanwhile unbelief is rising. How could it not?"[5]*

The Evolution of God and Religion

With increasing numbers of people worldwide already walking away from traditional religions, new religious doctrines and a new definition of God will have to emerge for any religion to survive. And for those who might argue that our interpretation of God can't change, consider the dramatic about-face in the description of the Old Testament God who was misogynistic, jealous, cruel, vengeful, sexist, etc. to that of the New Testament God who suddenly became loving, compassionate, and forgiving (although by the end of the New Testament, God had returned to his old vengeful, jealous self as he sentences non-believers to an everlasting burning pit of fire). In the Christian Bible (from the writings of the Old through to the New Testament) we see clear evidence of the evolution of God, and this evolution will continue in the 21st century if any God-based religions are to survive.

As bestselling and award-winning author, Robert Wright, writes in his book, *The Evolution of God*:

> *"Religion needs to mature more if the world is going to survive in good shape—and for that matter if religion is to hold the respect of intellectually critical people."[6]*

When Wright uses the word "mature" in the previous quote, he is referring to how both religion and our perception of God have evolved throughout our history, and how this evolution needs to continue in the 21st century if religion is to hold any validity.

In terms of religious beliefs today, many people can't see beyond the so-called Holy Scriptures of the past, books that are sometimes out of line with modern societal trends and morals. Lawrence M. Krauss, director of the Origins Project at Arizona State University and Foundation Professor in the School of Earth and Space Explorations, writes:

WHY ARE WE HERE? / 247

"As a guide for understanding the world, the Bible is pathetically inconsistent and outdated. And one might legitimately argue that as a guide for human behavior large swaths of it border on the obscene."[7]

Perhaps there are books that have been written far more recently and books that remain to be written, that don't claim by their authors to have been inspired by God appearing to them in some kind of mystical manner, that will be superior guides to helping us find more appropriate and lasting meaning in the 21st century and beyond.

The religions that may prosper will be those that develop new doctrines and traditions that are more acceptable to people of the 21st century. Such doctrines will most certainly be more inclusive of all humans (giving equal rights regardless of sexual orientation and gender, as an example, as opposed to both Christianity and Islam that currently give greater rights to males).

What Will Be Among the First Religious Beliefs that Will Change?

If we look at the research related to current religious beliefs, we find that the doctrine of eternal damnation (that many religions maintain) is one of the least supported doctrines. A 2007 Pew Research study found that only 58% of U.S. adults believe in hell, but if we look at this a little closer, we find that only 27% of people, who do not subscribe to a particular religion, believe in hell.[8] In looking at other countries, including Germany, Norway, Great Britain, Australia, Denmark, Sweden, Canada, Spain, France, Japan and Belgium, it was found (based on interviews with 29,133 interviews in 38 countries) that a majority of people do not believe in hell (or heaven). As an example, only 19% of the people in Spain believe in hell.[9] A survey conducted by St. Andrews University found that "even a new generation of more theologically conservative ministers find it (hell) deeply unattractive."[10]

Take away the fear of hell and many religions would collapse.

Throughout history major religions often appealed to the common people because of the promise of a beautiful everlasting heaven (in comparison to their short, painful existence here on earth).

And when the belief in heaven wasn't convincing enough to gain followers, the concept of hell was a powerful way for church leaders to keep their membership obedient. The concept of hell also provided the illiterate masses with a source of satisfaction to know that the people who made their lives miserable here on earth would someday face eternal damnation in a fiery hell. The doctrine of an everlasting burning hell, in many churches, is one of the most disturbing perverse myths that has ever been promulgated in the name of religion.

As peoples' attitudes are changing in the 21st century regarding heaven and hell, how will this shape the religions of the future? And relevant to this book, how will these evolving beliefs impact the answer to the question as to why we are here?

While hell and eternal damnation may have frightened people into attending church in the past, this approach is less likely to work in the 21st century. Some churches will continue to promote such a doctrine and see their memberships dwindle while other churches will begin to downplay eternal damnation (finding some way to look at it more metamorphically or perhaps linking it to a past historical context, rather than the present).

It is very likely that a hundred years from now people will look back at the major religions of today in the same manner that most people now dismiss the pagan beliefs of ancient civilizations (even though if they explored the history of their religion, they would likely find its roots in the very pagan religions that they think are outdated). Religion is a human invention that meets the psychological needs of individuals and is acceptable to the cultural norms of society. As these needs and norms change, religions will either evolve or die.

Harry Potter Leads the Way

Although there are a growing number of atheists and agnostics in the world (more on this in a few more pages), the need to believe in the supernatural appears to be part of the fabric of many humans although this belief does not necessarily have to be expressed in some form of a traditional God.

For example, the incredible success of the *Harry Potter* series demonstrates the interest of many in the supernatural. I don't think

WHY ARE WE HERE? / 249

it would be an exaggeration at all to state that there have likely been more people in the 21st century who have read at least one Harry Potter book than who have read the Christian Bible (maybe even more than who have read at least one brief book in the Bible).

Interest and curiosity in the supernatural is not limited to Harry Potter. Look at any listing of TV shows or movies, or browse through books in a local bookstore (especially in the Young Adult section, and these are our shapers of religions of the future) and you will find endless titles whose content relates to the supernatural.

In an article on the origins of religions, Elizabeth Palermo, the associate editor of Live Science, writes:

> *"This tendency to explain the natural world through the existence of beings with supernatural powers—things like gods, ancestral spirits, goblins and fairies—formed much of religious beliefs, according to many cognitive scientists."*[11]

In many ways with the emergence of the fascination in the supernatural by many young people, we have come full-circle back to the pagan beliefs of our earliest ancestors. Religion appears to be ingrained in us, even if more and more people are rejecting the current teachings of the major religions in the world and are looking for something different than what they offer.

The Evolution of Church Doctrines and Practices

While some might argue that churches will never change their doctrines or practices, such a statement is simply not true. Many religions have changed their beliefs throughout their history. Like everything else, religious beliefs have evolved. The law of natural selection impacts religious beliefs just as much as it does biological change.

For example, recently the Pope said that the death penalty is an attack against the dignity of all humans. He argued that the Catholic Church must move towards abolishing support for capital punishment, even though capital punishment has been a longstanding belief of the church. The Pope said that the previous belief was outdated. This is an example of a belief within a major church evolving to better appease its membership and be more in line with societal thinking.¹²

The Catholic Church no longer burns people at the stake if they don't believe in its teachings. It no longer skins non-believers alive or stones them to death or stretches them on a rack until they split apart. These are but a few instances of a religion that has altered its practices to better fit into evolving societal norms.

We might also ask how increasing sexual abuse claims against priests in the Catholic Church will impact people's views about Christianity. In August of 2018 a U.S. grand jury report revealed that more than 1,000 young boys and girls were molested by some 300 Catholic priests since the 1940s in the state of Pennsylvania.¹³ Imagine how horrendous these numbers would be if all 50 U.S. states were considered or all the countries in the world where the Catholic religion is found.

These ugly numbers are not an aberration. They are not isolated to the 20th or 21st century. If there was a way to go back in time, we would find a disturbing trail of such abuse going back to the beginnings of the Catholic Church.

Lest Protestants feel a little smug about their own faith, perhaps you remember some statistics that were presented in Chapter 6 of this book that showed that some leaders within the Protestant church also have a dismal record of abusing children as well.

And let us never forget that if it wasn't for the horrendous persecution of non-believers by the Catholic Church throughout history, it's unlikely that any form of Christianity would still be with us. Christianity is built on fear and violence. As statues of past leaders who were involved in slavery and other crimes against humanity are being torn down, how long will it be before religious icons are dismantled because of the historic horrors that they represent?

Change is coming.

When I was a member of an evangelical Christian church decades ago, most adherents believed in the literal interpretation of Biblical stories such as Adam and Eve, or Noah and the flood. Today, many members of this same church (even including some leaders within the church) believe that such stories are metaphors rather than the literal truth. This is another example of religious beliefs evolving to be more in line with societal norms (which in this case have also been strongly influenced by science and archaeology which do not support the original Bible stories).

Even religions like Hinduism which are very heavily steeped in traditions are changing to parallel what is happening in society rather than remaining glued to past teachings. For example, recently Indian women have been permitted to enter the historic Sabarimala temple (which sees more than 50 million devoted visitors each year). Previously women from the age of 10 to 50 had been barred from entering the temple because in the Hindu religion it was thought that when women menstruated they were impure. In the Hindu religion it was taught that the Hindu god Kartikeya curses women who enter this temple (as well as other Hindu temples). Chief Justice Dipak Misra, of India's Supreme Court, stated: "Religion cannot be the cover to deny women the right to worship."[14] As society norms and laws change, religion will follow.

Our interpretation of God and church doctrines will continue to evolve as they have throughout history. One can only hope that as the future face of God evolves that this power, or force, or source of energy (or whatever descriptors are used) will become a unifying force, rather than a source of division. How different would our world be if the God of the 21st century inspired peace instead of conflict, and a belief in this God resulted in concern for the basic needs and rights of everyone who inhabits our planet?

Finding Meaning in Religion

For readers who are a member of one of the thousands of religions that exist in our world, if your religion offers a meaningful answer to you in response to why we are here, that's great. The only caveat I would add is that religious people need to accept others regardless of their faith (or lack of faith). Our world will never find peace if one religion believes it is superior to another religion, or if one religion sees its mission as saving anyone who doesn't believe in its teachings.

When you have members of Jehovah's Witnesses or the Mormons knocking on your door, or some Muslims standing on street corners handing out free copies of the Quran, or church congregations still singing songs such as *Onward Christian Soldiers,* this is visible evidence that we still live in a world where some religious groups have a major goal of converting others to their religion (which inherently means that they believe that their religion is superior to other religions). And don't think for a moment that members of evangelical churches don't want to bring you into their fold even if their methods might not be so obvious as knocking on your door.

At a time of nuclear weapons and other new weapons of mass destruction, our species cannot afford to be the pawns of religious leaders as they attempt to convert unbelievers to their antiquated views.

My wife and I recently attended a wedding where the bride was Punjabi (and her family were members of the Sikh religion) and the groom was Canadian (and his family were mostly members of the Christian religion). When we arrived at the temple where the wedding was being held, the groom's family gathered at one end of the parking lot while the bride's family gathered at the other end. At a designated time, the groom's family walked towards the bride's family (I was told that in India, the groom often rides a white horse at this moment).

When the two families converged, several members from the families embraced and placed a garland of flowers (like a Hawaiian lai) over their heads. This act was symbolic of the two families coming together, regardless of their cultural and religious background. This is the kind of acceptance religions need to embrace around the world. Members of one religion need to share garlands of flowers, not guns or propaganda, with members of other religions.

I believe that many millennials are already far more accepting than their parents in accepting those from other religions. This is a trend that could most definitely contribute further to world peace. As my daughters grew up, I often found it interesting that at Christmas, they shared gifts with their friends, many of whom were Muslim, Jewish, or other religions, or even non-believers. In their eyes, Christmas was viewed as a time of giving to others, regardless of their religious background. Our young will usher in new traditions and from what I've seen, this could be a positive change for humanity.

In looking at the acceptance of another's religious beliefs, there's also a need for believers to accept non-believers, and for those who are atheists or agnostics to accept those who still believe in God and find life's meaning in the teachings of their religions. Acceptance must be universal, provided that religious teachings are not being used as a basis to persecute others in any manner.

Regardless of one's religion, there is a need to focus on living a quality of life today, instead of always looking forward to life after death, a belief that in fact may or may not be true. How would our world change if everyone was focussed on living the best life now instead of thinking that the best is yet to come in heaven? As John Lennon wrote, "Imagine there's no heaven."

If Religion Doesn't Help You Find Meaning

If your religious beliefs do not provide you with a meaningful or acceptable answer as to why you here, then perhaps one of the answers provided in the remainder of this chapter might be helpful to you.

A New Definition of God

Previously in this chapter, I mentioned that our definition or description of God is evolving as it always has. It's even possible that the word God itself (which has had so many negative connotations throughout history) may be replaced. Instead of speaking of God, some have replaced this name with words such as Force (more specifically known as Jediism), Power, Supreme Being, Energy, and even Spirit (but a different spirit than the traditional spirit associated with God).

There is even a Church of Google (with their own Ten Commandments and official prayers). And there are some new religions with more out-of-the-box definitions for God such as the Church of the Flying Spaghetti Monster (Pastafarianism) and the Church of Cannabis.

While these latter churches may never rival the more traditional religions, let's return back to Jediism with its roots in Star Wars. Across the world there are people on regular census surveys who identify their religion as Jediism. In Australia, as an example, there are currently at least 50,000 people who were identified through government census results as being a member of the Jedi order. In New Zealand there are more Jedis than Hindus or Buddhists. Robyn Faith Walsh, a professor of Religious Studies, writes:

> *"It took generations after the death of Jesus, for example, for Christianity to develop into something we might recognize as an established religion. So, one could argue that it's plausible that that a beloved narrative like Star Wars could develop into a more concrete set of practices and beliefs."*[15]

May the Force Be With You

The concept that God may be more a force of energy than a father figure with a white beard has a growing number of adherents in the 21st century although most of the believers of such a force wouldn't describe themselves as being part of a religion. There are a number of factors that are contributing to this belief, often based on scientific facts that have been misinterpreted beyond their primary significance.

For example, the first law of thermodynamics (related to the law of conservation of energy, in this case heat) states that the total energy of an isolated system can neither be created or destroyed. Some people go beyond the intended definition of this scientific law and use it to suggest that humans (who could be said to have an isolated energy system) can not have this energy system destroyed upon death which is then interpreted that our energy system remains for eternity. Of course, there is absolutely no proof that such an energy system has any form of consciousness, although a few new scientific theories have been interpreted by some as supporting the idea that consciousness may be found in the smallest particles that exist throughout our universe.

At the Large Hadron Collider (LHC) at CERN, in 2012, scientists confirmed the existence of a very small particle that is generally known as the Higgs boson, but has been nicknamed by some as the "God Particle." The Higgs boson describes a fundamental force existing throughout our universe that is transmitted by particles known as gauge bosons. There are some who believe, without scientific proof (but then again aren't most religions based on faith because they lack any scientific proof) that the Higgs boson may be the very fabric of God.

The Matrix

Add to this the String Theory which "proposes that all objects in the universe are composed of vibrating filaments (strings) and membranes (branes) of energy."[16] To explain String Theory, it is necessary to have 10 dimensions (six of which we currently can't view in addition to our three dimensions of space and our one dimension of time). If it is true that we live in a universe where there are an additional six dimensions that we can't view, is it possible that there are a vast number of universes tucked away within our universe, and if so, how might this affect our thoughts about why we are here?

Universal Consciousness

Some people will interpret these scientific findings (specifically the Higgs boson and the String Theory) to prove that God exists in the tiniest of particles throughout our universe and that as we are all joined together through these small particles (and fields) that this might somehow create a universal consciousness. Some have suggested that we are all part of one larger body (and that this body cannot be destroyed). As Dr. Gerald L. Schroeder, a MIT-trained scientist, writes in his book *The Hidden Face of God*,

> *"A single consciousness, an all-encompassing wisdom, pervades the universe. The discoveries of science, those that search the quantum nature of subatomic matter, those that explore the molecular complexity of biology, and those that probe the brain/mind interface, have moved us to the brink of a startling realization: all existence is the expression of this wisdom. Every particle, every being, from atom to human, appears to have within it a level of information, of conscious wisdom."*[17]

How such scientific findings might contribute to the creation of actual religions or contribute specifically to one's search for meaning remains to be seen, although in many ways the concept of spirituality which is a very popular trend, especially among younger people, may use such findings to support their beliefs.

A Growing Trend Towards Spirituality

Spirituality (not the same as spiritualists) believes that we are all somehow connected to each other (which means we should be kind to each other) and to the universe (which means we should care for our planet) and that we can all seek meaning in our own individual way.

While a spiritual person may seek an experience that could be considered sacred, this experience is not necessarily defined by Holy Scriptures or the doctrines of a church. In fact, many spiritual people tend to reject anything to do with a church. Another component of spirituality that is attractive to many younger people is the emphasis on self-growth which can include getting in touch with one's inner dimension.

A 2012 Forum Research poll in Canada showed that two-thirds of Canadians identify themselves as being spiritual while just half identify themselves as religious.[18]

Spirituality allows people to give their own meaning as to why we are here or what happens to us after death. It is this personal interpretation related to the big questions that is attractive to growing numbers of people. Spirituality allows a person to get rid of all the harmful problems of major religions that have been outlined throughout this book while still believing in a God or a Force or a Power.

Spirituality allows a person to put their own spin on eternal life (and what it might look like) as well as personally interpreting if, and how, God might be involved in their life. After talking to a number of young people who believe in spirituality, I think it is only a matter of time before someone takes the core of these beliefs and establishes a religion that is free from some of the problems of past religions, a religion that is more encompassing and accepting of individual interpretations of God, as well as being more considerate of all life on our planet.

What is the Fastest Growing Religion?

Although not actually a religion, atheists (those who don't believe in God or gods) and agnostics (those who aren't sure whether there is a God or not) are among the largest growing group on our planet. This number is rapidly increasing as many mainstream church memberships are dwindling. It is estimated that there are currently 450–500 million atheists and agnostics in the world (about 7% of the world's population). If we expand this further to include non-religious people in general, according to the Pew Research Center this number then jumps to 1.1 billion people which is about 16.5% of the adult population.[19]

It becomes even more interesting when we look at individual countries. It is reported that 40-49.9% of China's population identify as having an agnostic tendency while between 30-39% of Japanese claim that they are "convinced atheists." The same percentage holds true for the Czech Republic and 20% of French citizens state that they are "convinced atheists." In the U.S. twice as many Americans said in 2014 that they did not believe in God as compared to similar research done in 1980.[20] During the 21st century is it possible that atheists and agnostics could outnumber those who believe in God? If so, how might this change our world?

As reported in *National Geographic*, the religiously unaffiliated are the second largest religious group in North America and most of Europe. This article in National Geographic stated that the religiously unaffiliated in the U.S. have overtaken Catholics, mainline Protestants, and all follower of non-Christian faiths. Gabe Bullard, the author of this article, states:

> *"There have long been predictions that religion would fade from relevancy as the world modernizes, but all the recent surveys are finding that it's happening startlingly fast. Religion is rapidly becoming less important than it's ever been, even to people who live in countries where faith has affected everything from rulers to borders to architecture."*[21]

How Do Non-Religious People Find Meaning in Life?

There are some who would ask how an atheist/agnostic can find answers as to why we are here if they don't believe in God. In reviewing a new book titled *A Better Life: 100 Atheists Speak Out on Joy and Meaning in a World Without God*, David Niose writes:

> *"A common theme emanating from this book is that*

> atheists have little trouble finding purpose in their lives. Having rejected myth and ancient texts as authorities for defining life's purpose, nonbelievers get meaning and joy from family, friends, loved ones, nature, art, music, and their work."[22]

When a person no longer believes in an afterlife (at least as described or suggested in various Scriptures), they often live a more vibrant life each day. If you go through life thinking that the best is yet to come when you receive your reward in heaven, you might very well fail to realize that the best is available to you right now, every minute of every day. Of course, some religious people will say that their personal relationship with God enriches their lives each day, although I have yet to meet a religious person that had some vibrant intangible quality that was any different than an optimistic non-believer.

During the past few years my mother and my mother-in-law moved into long-term care facilities. Both women could definitely be described as being religious, although they were members of two different Christian denominations. During their final months and days, I never heard either women talk about her desire to get to heaven. What I did see was their faces light up when family members visited them, even when they sometimes struggled to remember names. What I also heard was a desire to die, a desire to end their struggles, a desire to end what they both viewed as non-quality living. This desire to die was never further elaborated with the words "and then I'll go to heaven."

At the end of a long, but fulfilling day, you might drop into bed exhausted and readily welcome sleep. Considering that sleep, apart from dreaming, is a time of nothingness (which most of us look forward to at the end of a day), would it be so bad, so depressing, so meaningless, if at the end of our lives there was nothing else? For those who suffer in pain, nothingness would mean that pain no longer existed. For others, the experience would be similar to falling asleep except that it would be final. And for those who lived every day to its fullest, they would fall into their final sleep filled with happiness and satisfaction.

Would there still be tears and grief? Of course. It's normal for

WHY ARE WE HERE? / 259

most individuals if they know they are facing death to be sad to say goodbye to loved ones (or even fearful of dying). It's also normal that any loved ones would go through various forms of grief when they have to say their goodbyes. And I fully acknowledge how important the belief in an afterlife can be in grieving loved ones who have died, especially if the death was the result of a tragic event, and particularly if the death involved a child, but as I suggested back in Chapter 5, most people put their own spin on what the afterlife really looks like, and this interpretation is generally significantly different (and more meaningful to those who are grieving) than what we find in most Scriptures.

If an atheist or agnostic chooses to believe in nothingness when the death of a loved one occurs, then the loved one can remain with them in their memories. Such a scenario is not depressing. Such a scenario is not terrifying. Such a scenario recognizes the realities of our finite physical bodies. Such beliefs can even contribute to a more meaningful and vibrant life, living in the present moment instead of a vague promise of a better future which has no factual basis. Is it better to die having lived the best life possible, rather than die having avoided ever living?

As we encounter more atheists and agnostics in our society, we will see a change in funeral practices. It will no longer be necessary to talk about sin, or hell, or even heaven. Funerals will focus more and more on celebrating the life of the deceased instead of focusing on the hereafter. This trend is already happening, even when the deceased is a devout member of a church.

Adam Lee, in an article about facing death without religion, writes:

> *"Instead of following the script that's been written for us,*
> *we can create our own customs and choose for*
> *ourselves how we want to be remembered. We can*
> *design funerals that emphasize the good we did, the*
> *moments that made our lives meaningful*
> *and the lessons we'd like to pass on."*[23]

A new study from Coventry University, found that the most religious and atheists both had the least fear of death.[24] The point here is that both believers and non-believers can achieve their own meaning in life and in death, even if both groups find it in different ways. In the 21st century we need to recognize and accept such differences.

The late Stephen Hawking, often considered to be the most world-renowned scientist since Einstein, said:

> *"We are each free to believe what we want, and it's my view that the simplest explanation is that there is no God. No one created the universe and no one directs our fate. This leads me to a profound realization: there is probably no heaven and afterlife either. I think belief in an afterlife is just wishful thinking. There is no reliable evidence for it, and it flies in the face of everything we know in science. I think that when we die we return to dust. But there's a sense in which we live on, in our influence, and in our genes that we pass on to our children. We have this one life to appreciate the grand design of the universe, and for that I am extremely grateful."*[25]

Acceptance of Non-Believers

Unfortunately, the biggest worry that atheists face doesn't come from the judgment of a god after death; it comes from religions here on earth that are unable to accept others who have different beliefs. A study in 2013 by the International Humanist and Ethical Union found:

> *"In 13 countries around the world, all of them Muslim, people who openly espouse atheism or reject the official state religion of Islam face execution under the law. And beyond Islamic nations, even some of the West's apparently most democratic governments at best discriminate against citizens who have no belief in a god and at worst can jail them for offenses dubbed blasphemy."*[26]

WHY ARE WE HERE? / 261

When Barack Obama was elected President of the U.S. this was considered a historical first because of his African American heritage. When a woman becomes President of the U.S., this will be considered another historical first. Imagine how times will have changed when an atheist becomes President of the U.S. What words will replace "God Bless America" at the end of every speech that this president gives?

Let's look at some other possibilities related to answering the question as to why we are here that have emerged in the 21st century.

Breakthrough Starship

In our search for life beyond our planet, one of the things we have done is to send satellites on interplanetary probes to other planets in our solar system and even to the farthest-reaching extremities of our universe. Voyager 1 and 2, both launched in 1977, have now travelled to the distant edge of our solar system and are now breaching interstellar space. Both Voyager spacecraft carry 12-inch, gold-plated copper disks with recorded messages from us in the event that another intelligent lifeform discovers the spacecraft.

How long will it be before instead of sending scientific instruments to explore distant planets or recordings to introduce us to other possible advanced civilizations, that we begin to send life itself beyond our planet?

While we have not reached the technological knowhow to send people on space journeys of hundreds or thousands of years similar to Arthur Clarke's *Space Odyssey 2001*, we could send the very fabric of the beginnings of life as it occurred on our planet. We could send our most primitive lifeforms that upon impact on some distant planet (or even an asteroid that might randomly deliver them to some distant planet) might create a new form of evolution on that planet. This idea could speed up dramatically if we reach the global warming tipping point and humans face extinction.

A project known as Breakthrough Starship is already in its planning stages. This project funded by a Russian billionaire and scientists from around the globe hopes to send miniature spacecraft throughout our universe, each filled with microbes that could seed new life on distant planets.[27]

There may be ethical issues in doing this, but that's a discussion for another book. I introduce such a venture only for readers to consider the opposite possibility. Instead of humans sending basic lifeforms out into the universe, what if some other advanced civilizations have already done this and our planet (billions of years ago) was the recipient of such a venture?

Transpermia

Transpermia (the word panspermia is also used) is the hypothesis that basic life forms that exist throughout our universe are carried by space dust, meteoroids, asteroids, and even space ships and deposited unknowingly on other planets. While panspermia is defined as a naturally occurring random phenomenon, it's also possible that an advanced civilization somewhere in our universe sent out their own versions of Voyageur spacecraft filled with basic lifeforms (or the chemicals necessary to begin life) billions of years ago.

Life as it began on our planet may have in fact come from some other planet in our universe, whether randomly or by design. Within those basic lifeforms could have been the blueprint for the process of natural selection. Instead of a God creating us, we may be the descendants of life from some other planet.

How would such a possibility impact our search for meaning in the 21st century? Such a possibility would preclude the claim of most religions that humans were uniquely created by God. Such a possibility would also tell us that we are not alone and we may certainly not be the most advanced lifeform in our universe.

If scientists could prove that life on our planet originated elsewhere in our universe, or that through transpermia there may be lifeforms developing on our planet right now that were not part of the original creation of life on earth billions of years ago, this would rattle many of our beliefs about why we are here.

Are We Living in a Virtual Reality?

In the *Introduction* of this book, the possibility that we might be living in a virtual reality simulation was mentioned which provides another theory about why we are here.

A month ago, my family and I visited a virtual reality gaming center where we each placed on our virtual reality goggles and en-

WHY ARE WE HERE? / 263

tered various imaginary worlds, although some of these worlds were so real that the line between reality and illusion became blurred.

As a first timer to a VR gaming room, I had been told by others that one of the "must" experiences was "walking the plank" which likely goes by a variety of names in different VR programs, but the intent is still the same.

In the version that I experienced, I stood at the beginning of a wooden plank (think of a narrow diving board) that stretched out from the top of a very tall building over a cityscape. The sense of height was so real that I hesitated to move. Finally convincing myself that the city streets far below me were not real, I gingerly stepped forward along the plank, trying to keep my balance lest I fall off the plank and virtually plunge to the pavement on the street far below me.

Finally making it to the end of the plank, there was an option to step off the plank and fly like a bird. The illusion was so real that I didn't dare step off the plank. The overall experience was so lifelike that it is apparently not unusual for some people to lose their balance and fall off the plank, only to find themselves literally sprawled on the floor of the gaming room after they remove their VR goggles.

In chapter 8, we looked at how advances in VR are reshaping education, entertainment, medicine, and even the treatment of the elderly. As VR continues to advance, the worlds that will be created will be beyond belief (pun intended). In a few short years, it's likely that the VR advances of tomorrow will make today's VR look like the primitive "Pong" game that ushered in digital gaming decades ago. Will we soon be entering the virtual worlds of current fantasies such as *Ready Player One*?

One area of VR that we didn't look at though in Chapter 8 was the possibility that we might actually be living in a VR world right now. Is it possible that our world isn't real, at least not in the sense we think it is? While many might scoff at such an idea, consider our own bodies. It's estimated that our bodies are comprised of a huge number of atoms (7×10^{27} to be a little more exact). Atoms are comprised of 99.9999999999996% empty space. As a result, our bodies are basically empty space (as is anything that we can see), but our eyes and brains work together to give us the perception of actually seeing other people or things. Is it possible that what we think we are seeing is only an illusion? Is it possible that in the same way that digital pixels are used to fabricate images in a VR world, someone or something has used atoms to create our world?

Could We Be Living in A Virtual World?

A few weeks ago, I read an article that suggested our lives were nothing more than a giant virtual reality game that someone from another world has created. Over the following weeks, similar articles popped up in multiple publications. Elon Musk (the CEO of TESLA, the founder of SpaceX, and a man who is considered to be one of the great minds of the current technological revolution) was reported to have said:

> *"We are simulations living in a virtual realm."*[28]

While some might argue that the 17th century philosopher René Decartes hinted at such a possibility when he talked about the knowledge we can gain from sensory illusions and dreams, he didn't actually use the term virtual reality (because it didn't exist).[29] And it could be argued that such thinking can be traced all the way back to Plato in his Allegory of the Cave.[30]

The growing interest in the possibility that we might be living in a virtual world is directly related to advancing technology. Virtual reality games now exist where imaginary worlds can be created and characters can be manipulated. These games have been a huge factor in contributing to both the resulting new vocabulary and technological advances related to virtual reality. What was implausible a few short years ago could now be (sorry) reality. Mark Zuckerberg, the founder of Facebook, said:

> *"Virtual reality was once the dream of science fiction. But the Internet was also once a dream, and so were computers and smartphones. The future is coming and we have a chance to build it together."*[31]

Julian Baggini, a British philosopher, writes:

"Human beings have long been fascinated by the idea that the world as it appears to us is not the ultimate reality. In recent years, however, such metaphysical speculations have taken on a more materially conceivable form. Computer-based virtual reality makes the idea that we could be living in a simulation more than just an abstract possibility; some very smart people even think that this is not only possible, but likely. Very likely."[32]

Are We Living a Dream?

For anyone who has ever had a vibrant life-like dream (which is likely just about every reader), as real as any dreams might seem to be, they are illusions created in our brains. Is it possible that the world in which we live is a dream created by a massive consciousness that permeates our world? The movie *Inception* explores the possibility of consciously entering our dreams.

As Brian Greene, one of the leading physicists in the world and a Pulitzer Prize finalist, writes:

> *"What is reality? We humans only have access to the internal experiences of perception and thought, so how can we be sure they truly reflect an external world? Philosophers have long recognized this problem. Filmmakers have popularized it through storylines involving artificial worlds, generated by finely tuned neurological stimulation that exist solely within the minds of their protagonists. And physicists such as myself are acutely aware that the reality we observe—matter evolving on the stage of space and time—may have little to do with reality, if any, that's out there."*[33]

In the same manner that we can't prove that God created our world, we can't find evidence (at least not at this time), that our world is only a VR simulation created by some superior being or force. Similarly, at this time we can't prove that we might also exist in another dimension. If our world is really a dream or a VR experience, how would this change our thinking about the meaning of why we are here?

Are We Also Alive in Another Dimension?

Is it also possible that there may other dimensions that surround us? We generally accept that we live in a 3-D world but could there be more unseen dimensions? In a 2016 article in *Universe Today*, we read:

> *"Dimensions are simply the different facets of what we perceive to be reality. We are immediately aware of the three dimensions that surround us on a daily basis...Beyond these three visible dimensions, scientists believe that there may be many more. In fact, the theoretical framework of Superstring Theory posits that the universe exists in ten different dimensions. These different aspects are what govern the universe, the fundamental forces of nature, and all the elementary particles contained within."*[34]

Even if the 10 different dimensions—suggested by the Superstring Theory—gains proof, this does not necessarily mean that we also exist in some other dimension. Even if we are living in a virtual world or there might be such a thing as a parallel universe, we can still only live in the present moment that we have.

To live one's life for something unforeseen in the future can prevent one from living life at all. Regardless of how our world started or regardless of whether other dimensions might exist, or even regardless of the possibility that we might be living in a VR world, we all have what we wake up to in the morning. We can't change the past and we can't accurately predict every element of our future, but we can find meaning in what we do today. This is our reality regardless of its source. We can choose how we give this reality meaning.

AI and Humankind

In looking at finding meaning in the 21st century, we might also ask what role artificial intelligence (AI) is going to play. Recently a brothel of artificial lifelike silicon dolls opened in Toronto (and closed just as quickly due to a neighborhood outcry), but it's only a matter of time before it reopens.

While there is something significant about a brothel opening with robotic dolls (if they could even be called robotic because according to newspaper reports, the dolls did not move or speak, although the next generation of these dolls will most certainly do this), there is something even more significant about a community protesting against their presence.

Are people saying that the laws that impact human morality must also be applied to silicone dolls? If so, will we soon be granting rights to a multitude of AI devices (think Siri, Google Home, Amazon Echo, etc. as examples of an ever-expanding field)? Will people legally marry robots? Will robots be considered legal entities?

In Love with an AI Doll

I recently watched a documentary that showed some people in Japan having a relationship with an online imaginary person who through AI was able to provide a meaningful conversation (and the same companies such as Sony and Hitachi that are developing these online companions are also looking at enhancing the experience by adding holograms).

Some people take their online AI companions to bars or other social functions. In addition, the documentary showed lifelike silicon dolls that could speak and who had some limited mobility. If some people find a meaningful relationship with an online avatar, how much more satisfaction will they experience with a lifelike doll that may cater to some of their needs?

Vinclu, a Japanese company, is currently taking orders for an interactive anime character that appears as a hologram in a glass tube about the size of a coffee-maker. This device known as Gatebox takes Alexa or Google Home to a new level.

Soon, other companies will provide this product at an even more advanced level, and it won't be long before this virtual image can carry on conversations well beyond the normal, "What is the weather going to be?" or "Turn on the lights."

Not only can a device such as Gatebox provide a virtual image in the home, the presence and functions of this image can be added to one's cell phone and provide ongoing companionship throughout the day for a user.

Finding Meaning in a Relationship with a Fantasy Image

Relationships can have a huge impact on providing meaning for us. This has always been one of the advantages of religion. Belonging to a religion often helps a person to connect with other people with similar beliefs which can be beneficial to all involved. For a significant number of people, the answer to why we are here is found in a meaningful relationship with another person (or other people), and similarly for many people struggling to find meaning, their absence of meaning is often related to a failure to form or maintain significant relationships with others.

AI and VR will usher in a new era where anyone will be able to experience a meaningful relationship even if it is with an imaginary person. And given the number of cell phones worldwide, this will be an experience that will have the potential to stretch into the most remote regions of our planet.

When Computers Surpass Us in Intelligence

The 21st century is going to bring new technological advances that are beyond our wildest imagination. These advances may also bring new questions that we have never faced in our many years of existence as Homo sapiens. For example, once AI surpasses our own intelligence (as it most surely will), some might ask whether our purpose on this planet was simply to evolve to the point where we were able to create this super intelligence. As Elon Musk tweeted:

> *"Hope we're not just the biological boot loader for digital superintelligence. Unfortunately, that is increasingly probable."*[35]

The Internet as the New God

Earlier in this chapter I mentioned the Church of Google. The Internet is becoming both omnipresent and omniscient in the same way that God might be described (and keep in mind that the World Wide Web only had its beginning in 1990). What impact on our lives will it have in another 30 years, or even 300 years?) It is more than just a remote possibility that the God of the future may be found in some form of advancing technology.

For many people, technology is already more important than anything a church might offer. Recently I found myself in a church for a wedding, and it was staggering to me to see how many people were glued to their cell phones throughout the service.

Many schools are attempting to implement more technology while at the same time trying to solve the problem of students being more absorbed in their cell phones, iPads, and/or laptops, looking at content that is different than what is actually being taught in the classroom. And the curriculum at any level (or subject area) of our educational system (from kindergarten through college) doesn't come anywhere close to accepting that any student with a cell phone has instant access to content that they are often still being forced to memorize in meaningless drudgery.

So why are we here? In answering this question, while we might consider thoughts from the past that addressed this question, in the end we need to reframe our answer in light of what is happening at the time when we live, which just happens to be the 21st century.

Giving Meaning to Our Lives

Orchestrating the survival of every human, every religious thought, every political system, and everyone's search for meaning is natural selection. The most basic reason for our existence is the preservation of our species.

Fortunately, unlike most, if not all, other lifeforms on our planet, we can find meaning beyond the basic law of natural selection, beyond the survival instinct of the preservation of our species. Our minds allow each of us to find further meaning each day.

This chapter has provided an opportunity to consider some basic ways in which people can find meaning in the 21st century. This list of possibilities could be much longer.

While natural selection may be the driving force of our existence, in the latter part of this book it has been argued that cooperation may be our greatest ally in the future survival of our species (and of our planet). As such, we can perhaps find our greatest reason for existing through our relationships with others. We are a social animal (not the only one) who can achieve significant meaning through practical kindness and compassion to others (and to ourselves). Every day we can find meaning through our connections with others, regardless of our religion, political beliefs, sex, race, or ethnic origins. Added to this are people who find companionship and meaning through their pets.

If global warming is a reality (and it certainly appears that it is), we are facing the greatest peril since the beginnings of our species, a peril that is much greater than any wars of the past. This peril threatens the survival of our species and that of many other animals. Through a cooperative effort we can meet this challenge.

Through cooperation, we have the ability to solve other world problems such as disease, hunger, poverty, and inequality. The choice is ours. In our efforts to meet these challenges, we will find the answers to the why questions as we reach out to help others. True meaning can come from being a part of the movement to aid all of humanity in our survival and in our attempts to create a just world. The following provide some thoughts for us to consider as we look for meaning in the 21st century:

> *1. Life is what we have at this moment, not what we think we might have at some distant point in the future.*

> *2. Human rights are universal regardless of political or religious affiliations.*

> *3. Every individual should be free to seek his/her path to happiness provided that all human rights are respected in the process.*

> *4. Our planet is our home. We must treat it with respect.*

> *5. Being accepted is a universal right.*

> *6. Change is constant. Natural selection favors those who best adapt to change.*

> *7. Cooperation is more important to our future survival (and current state of happiness) than aggression.*

- 12 -
Some Final Thoughts

It is difficult to end a book about 21st century learning when we are only just ending the second decade of this century and change continues to be rampant. While Chapter 11 was the intended ending of this book, during the editing process I found it necessary to add a few more thoughts.

We live at a time of unprecedented change. A few hundred years ago it was possible for a person to have read all the new books that were published in any given year of that era. That has dramatically changed. According to data from UNESCO, there were 2,240,569 books published alone in 2010.[1] It is no longer possible for any person to keep abreast of everything that is being written, everything that is being researched, and everything that is being discussed around our globe. The most educated person can barely scratch the surface of what is happening in our world.

For most people, change is uncomfortable. It is far easier for many people to cling to old superstitions or traditions even if they have little factual basis and even if they can actually get in the way of being happy. Some people cling to destructive behavior even if it causes them pain because it is often more comfortable for them to hold on to inappropriate behavior or false beliefs than it is to embrace change. For those who accept change and attempt to learn more about at least some of the changes that are occurring, they can better contribute to the survival of our species, and to their own individual happiness.

The Rapid Pace of Change

In writing this book, on many occasions I would discover something in the news, in a magazine, on the Internet, or in a local newspaper that impacted what I had just written. This morning as I read the local newspaper during breakfast, I found three articles that related directly to the content of this book. During this past weekend I read five magazines (*Time*, *Newsweek*, *The Atlantic*, *New Yorker*, and *Bloomberg News*). From these five magazines I counted 14 articles that related directly to the content of this book. A few months from now when this book is published, there will be some content that will already be out of date. Such is the pace of change in the 21st century. As a result, you might find it helpful to use some of the concepts in this book as a springboard for your own further research and education, rather than considering them the final word on any topic.

Staying Abreast of Rapid Change
With the incredible pace of change in the 21st century, it's difficult for even the most motivated people to keep abreast of everything that is happening. Even the most curious or most learned will struggle to be up-to-date regarding all the changes that are occurring in science or technology.

Unfortunately, there are others who will simply turn a blind eye to change. Some will do this because they are already too busy trying to survive each day to be concerned about changes that are occurring around them. For these people, rapid advances in genetic engineering or nanotechnology, as examples, are too far removed from their daily existence for them to be concerned.

Others will avoid educating themselves about changes that are occurring because they feel powerless to do anything about them. And there will be others who struggle to intellectually grasp some of the advances that are occurring in science and technology.

A significant number of people will unfortunately simply continue to believe what their leaders (whether political or religious) tell them to believe. For some, it is far easier and more comfortable to let others tell them how to think.

We can't expect that everyone is going to get on board with attempting to understand new discoveries and new approaches to solving old problems. And we can't expect that everyone is going to make an attempt to consider the ethical implications of new advances in science and technology.

Given that there are huge masses of people (likely the majority

SOME FINAL THOUGHTS / 273

of people on our planet) who will remain largely unaware of current advances in science and technology (or problems facing us such as global warming and/or overpopulation), there is a need for leaders in our society to help others understand some of the implications of changes that are occurring in the 21st century so that all people can at least make more informed decisions. One of the first steps in doing this could be to revise the curriculum in our schools to reflect the realities of 21st century learning. For students in any country not to be exposed to natural selection or climate change as a legitimate part of their studies would be as irresponsible as still teaching that the earth is flat, or that the sun revolves around the earth, or even that mental illness is caused by the devil.

The Unequal Distribution of Wealth

Recently I have noticed a growing number of articles regarding the unequal distribution of wealth throughout the world. Although this is not a topic that I specifically addressed in this book, there are those who believe that the rich may create a future elite class of people who enjoy all the medical marvels related to health and longevity that the masses cannot afford to use.

In 2017, Credit Suisse stated that the richest 1% of the world's population own more than half the world's wealth (and this wealth inequality is rising upwards every year). Along the same lines, the world's richest 500 people saw their wealth increase by $1 trillion dollars during 2017.[2]

It could be argued that the super-rich have improved our world through job creation or new product development. Consider the number of jobs that have been created by a super-rich person such as Jeff Bezos, the founder and CEO of Amazon. His success has helped countless others by providing jobs as well as providing a cheaper consumer product. Or consider Bill Gates, the cofounder of Microsoft. Like Bezos, Gates is obscenely wealthy, but through his foundation Gates is helping millions of disadvantaged people around the world by exploring cures for diseases such as malaria or by helping impoverished people to have better access to fresh water.

The potential problem in the future related to such unequal wealth distribution across our globe is that some of the rich, unlike Gates, may use their money to enjoy the benefits of advancing technology in medicine as an example, while the middle and lower classes may not be able to afford these new advances. It is not beyond the realm of possibility that the super-rich could even create, through genetic engineering, an offshoot of Homo sapiens that leapfrogs the normal natural selection process to create a race that is biologically and intellectually superior to the rest of the human population. If this were to ever occur, this new super race of humans could eventually eradicate the rest of us in the same manner that Homo sapiens over thousands of years eliminated other forms of humans.

An analysis from the U.S. Social Security Administration found that "Life expectancy for 65-year-old-men in the top half of the earnings distribution has increased by five years. For those in the bottom half of the earning distribution, life expectancy has increased just over one year."[3] With increasing technological advances in the 21st century, this gap will widen as the rich will often be the only people who will be able to afford new medical techniques.

Genetic engineering and designer babies will definitely be a part of our future. Will the rich use genetic engineering and related technologies to create children who are superior to other kids? As I write this final chapter, within the past 24 hours several news sources have reported that a Chinese doctor has created three babies that were genetically engineered (this has yet to be properly verified).

Will governments (who generally play into the hands of their wealthy donors) support the development of new drugs or changing laws (such as the legalization of cannabis) that could contribute to an apathetic drug-happy general public, who simply don't care what the rich are doing?

Some studies have demonstrated that smoking cannabis makes people less motivated and taking marijuana can lead to individuals becoming withdrawn, lethargic and apathetic.[4] Have we entered an era where the wealthy won't have to worry about uprisings from the general public? All the rich need to do is simply ensure that the masses have their 100+ channels of TV, high speed Internet, compelling video games, alcohol, and legalized cannabis which could lead to the general public being too intoxicated with pleasure, after a day at their mundane jobs, to ever worry about what the wealthy are really doing.

What Can the Average Person Do?

Of course, you might very well be asking what can the average person possibly do in standing up to the rich? First of all (at least in democratic countries), we vote people into power. How could anyone have possibly believed that Trump, as a billionaire, would care about the average American? It should have been no surprise when the first major thing the Trump administration accomplished was the reduction of taxes for rich people like Trump himself. Sure, the plan was sold as a reduction of taxes for everyone, but there was no hiding the fact that the part of the tax reductions that impacted the average person would be phased out by the time the Trump presidency ended, but the reductions for the rich would continue forever.

While the average person cannot keep up with all the rapid changes that are occurring in the 21st century, we can all make a better effort to elect politicians who have a genuine interest and proven track record of helping others. When a population chooses celebrity over integrity for its leadership, then entertainment has already become Huxley's soma. When a public becomes so dumbed down that it elects a fool, then the public itself has become the fool. In a *Washington Examiner* interview recently, the award-winning documentary film maker Michael Moore said:

> *"Trump is our Frankenstein, and we are Dr. Frankenstein. We have helped to create a situation that has allowed us to end up with Trump. The dumbing down of our society through the media, the lack of education through poor schools, allows for a dumbed-down electorate, and for him to be able to actually get 63 million votes."*[5]

Almost 100 years ago, American writer H.L. Mencken said it slightly differently:

> *"As democracy is perfected, the office of president represents, more and more closely, the inner soul*

> *of the people. On some great and glorious day the plain folks of the land will reach their heart's desire at last and the White House will be adorned by a downright moron."*[6]

If we want a future world where we come together in peaceful cooperation to enjoy the technological advances that are coming our way, we need to develop a much better educated electorate, and we need to all recognize that every voice and every vote counts.

The Dangers of Progress

In writing this book, there were times when I was exploring technological and scientific advances that were so staggering that I couldn't help but wonder about the dangerous implications of some of these advances being misused.

For example, the various applications of genetic engineering are stunning, but they can also be disturbing. Applying the same principles that are currently being used to sterilize female mosquitos through genetic engineering (which could result in the eradication of malaria or zika-carrying mosquitos), it could be possible to target the genetic structure of any race or ethnic group of humans and cause their complete extermination within a generation or two. Imagine the resulting horror if any ruthless and ambitious dictator possessed such technology.

Once again, there is a need for people to educate themselves (or for leaders to better educate the public) on the potential application of new advances in science and technology and help to voice opinions on appropriate guidelines for the ethical use of such advances.

Human-Driven Evolution

Humans are already manipulating life on our planet. Juan Enriquez (Chair of the Genetics Advisory Council at Harvard Medical School) and Steve Gullans (a professor at Harvard Medical School for 18 years), wrote in their book *Evolving Ourselves:*

> *"What has already happened in terms of what lives and dies on this planet has already tipped so far toward unnatural selection and nonrandom mutation that*

SOME FINAL THOUGHTS / 277

it is not even really DarWin 2.0 but a new logic of evolution, one based on different principles and mechanisms. It is no longer just natural evolution but human-driven evolution."[7]

Human-driven evolution has given us the power of the gods we once worshipped. The important question will become who is going to control human-driven evolution?

Eugenics

Eugenics, which focuses on the selection and preservation of specific human traits, was adopted by Hitler although it was implemented in the United States and Canada before the time of the Nazis.

In 1907, proposed laws were being written in the U.S. for the sterilization of "inferior" people. More than 30,000 people were sterilized in 29 U.S. states between 1907 and 1939 in an attempt to cleanse society of weaker people (such as the mentally ill and intellectually handicapped).[8] Two provinces in Canada (British Columbia and Alberta) were also involved in eugenics in the 20th century. In Alberta, 4,785 people received compulsory sterilization between the years of 1928 and 1972.[9] In South Carolina in the U.S. up until 1977 social workers had the official power to recommend that people they felt were incapable of caring for children should be sterilized.

Following in the footsteps of these attempts to decide who should have children, the Nazis would eventually sterilize between 300,000 to 400,000 people.[10]

Some countries such as Bangladesh still practice sterilization of women and men to help control their booming population.

With a growing trend to test fetuses and even newborn babies for defects, who is going to decide who lives and who dies?

Peter Ward, a NASA astrobiologist, asked this question:

"We have directed the evolution of so many animals and plant species. Why not our own? Why wait for natural selection to do the job when we can do it faster and in ways beneficial to ourselves?"[11]

Who is going to decide what is beneficial and what is harmful?

The Wars of the Future

Throughout history, wars have tended to be won by the side that had the best combination of an abundance of manpower, better strategies, and more advanced weapons. In many ways, this changed with the bombing of Hiroshima and Nagasaki. Suddenly one bomb could do more damage than thousands of soldiers. And with the development of that one bomb, suddenly any country in the world, regardless of its size, could become a threat to world peace.

The wars of the future won't require armies in the traditional sense. The use of drones and various kinds of robots will eliminate soldiers. The use of cyberattacks might even replace bombs. A biological weapon could devastate the largest of countries.

Where major wars of the past were generally orchestrated by governments or religious leaders, it's now possible that a future major war could be instigated by some super-rich megalomaniac person (the end-of-the-world James Bond type of scenario that until now has remained in the realm of fiction).

What Happens when Our Governments No Longer Control Science or Technology?

In many ways this has already happened. When you have a major political party in the U.S. that denies global warming or wants to put the brakes on stem cell research, or who reject evolution, it would appear that the leadership and membership of such a party are so absorbed in the past that they are not able to understand the scope of both the problems and possible solutions that will come our way in the 21st century.

When governments fail to take the initiative in dealing with change, private industry will take over. Consider the American space program that for decades was operated through NASA. We have now entered an era of the privatization of space explorations with SpaceX (founded by Elon Musk, the CEO of Tesla) and Blue Origin (founded by Jeff Bezos, the founder of Amazon). The implications of such programs should be obvious. The most significant technological advances of the future are going to be in the hands of a few who are not necessarily part of any government. And when a government attempts to regulate any of the technological advances of the super-rich, these futuristic entrepreneurs can relocate to another country that has less regulations, or use their power and money to buy the government support that they need.

I'm not saying that Jeff Bezos or Elon Musk are doing anything wrong here; I'm simply trying to make the point that the super-rich have the means to drive and implement many future technological changes, and not every one of these people will have the moral character to use these advances to the betterment of all humanity.

Similarly, we could look at the potential problems of AI (artificial intelligence) or VR (virtual reality). We are increasingly moving towards a world where the average person (which includes the overwhelming majority of people on our planet) will live in a world where AI increasingly does their thinking for them and VR increasingly provides an escape from reality. The controllers of such technology (the super-rich) will have the means to infiltrate the minds and homes of the world masses, and to a large degree make them intellectually impotent.

Our Choice

As the 21st century moves along we have a choice: we can remain mired in the traditions of the past that tend to turn us into obedient sheep, or we can educate ourselves on the problems (climate change as an example) or any ethical implications of advancing technology.

In democratic countries, our votes can help to support progress when it is ethically responsible. On the other hand, our votes can help to stop the application of some technologies when they are harmful.

In non-democratic countries, social media can and will play a role of growing importance. Even in democratic countries, social media could play a huge role in giving the average person a voice, a voice that could make positive changes in our world.

But that voice must become educated. A uniformed knee-jerk reaction to change is no better and no different than evangelical Christians who ignore global warming because they think that Jesus is going to return at any minute.

Imagine if we could reach the evangelical Christians in the U.S.

and convince them that climate change is a real problem. Imagine if this huge base of American voters shed their sheep-like behavior and supported politicians who took climate change seriously. One way to do this would be to implement the facts related to climate change into the school curriculum across the U.S.

Technology can drive incredible changes that will bring increased health and prosperity to countless people around our planet, or it can be the foundation for incomprehensible destruction. By becoming more aware of technological advances, we can all play a role in providing guidelines for their use.

Walk Slowly

We live at a time of incredible change, yet the most important change that we all face is personal.

Recently I was on a walk with my wife and one of my daughters. During the walk, my daughter commented that I was walking slowly. As I thought about this further, I came to the conclusion that this was a compliment (although I'm sure that wasn't the intention).

As the pace of life in the 21st century increases, we are often caught up in the rush. It often seems like everyone is racing, often trying to juggle several tasks at the same time. Technological advances have contributed in some ways to this increasing pace of life as the Internet, as an example, provides instant information.

I frequently hear people say that they are expected to do more at work in less time (or more at home in less time). As I teach my university courses to educators, I hear teachers and counselors telling me that anxiety among students is at an all-time high because students are feeling overwhelmed with everything that is happening around them (which of course is brought to them instantly on their cell phones).

Change can cause stress. The rapid changes that are occurring in the 21st century, that come our way through 24/7 Internet headlines cause greater stress than what people might have experienced in previous times (although one could argue that a great depression and two world wars in the last century were hardly non-stressful).

While attempting to be more aware of some of the changes that are occurring around us, perhaps we all need to focus more on walking slowly.

SOME FINAL THOUGHTS / 281

Walking slowly allows us to observe the world as it exists around us. With each step we can note the changing color of the leaves on the trees or the emerging blooms of flowers in gardens throughout our neighborhoods. We can watch the changing skies. We can watch the children playing. And we can disconnect from technology as we walk.

For me, walking is a form of meditation, which in the end is something we could all benefit from, regardless of the type of meditation that we choose.

In several instances in this book, I have talked about things that were happening in my backyard (such as observing baby rabbits), so it should come as no surprise that there are times when I enjoy standing at a back window (or sitting in the backyard) just observing. Whether it's a new hibiscus bloom, the blowing of the leaves on the trees, a billowing cloud, a bird fluttering from one branch to another, or several squirrels chasing each other, it's at times like this that I am at one with the world, and it is from times like this that I am then able to return back to work and better focus on whatever I am trying to accomplish.

Regardless of our beliefs, we need to consider that all that we have for sure is the moment—the power of now. And when we slow down to appreciate this moment and we express gratitude within ourselves for being able to enjoy this moment, we can find the peace and contentment that eludes so many.

Meaning does not have to be rooted in complicated philosophies or lengthy outdated Holy Scriptures. Finding meaning can be achieved by walking slowly, by concentrating every once in a while on your breathing—realizing that you have been given the gift of life, however it originated.

Finding meaning in the 21st century does not have to be difficult. We live in amazing times where the standard of living is constantly increasing in comparison to what humans experienced in the past. We live at a time when technology can bring us instant access to new learning (often free) and can bring us into contact with other people anywhere in the world. Technology can enrich our lives if we moderate our attraction to it, instead of letting it control us. We live at a time of incredible opportunities—the key is whether one is willing to embrace the incredible changes around us or whether one decides to remain mired in the learning of the past.

Be the Change You Want to See in the World

While these words are often attributed to Barack Obama, some suggest they were used years earlier by Mahatma Gandhi. Actually, it is difficult to ascertain who is directly responsible for these words. It could be argued that they had their roots in the famous quote by Henry David Thoreau that is often expressed as "Live the life you've imagined" (although his actual quote says "...and endeavors to live the life which he has imagined") or in Gandhi's words that said, "As man changes his nature, so does the attitude of the world change towards him..."

The first recorded usage of this quote appears to be in 1974 (Ghandi died in 1947, so it's unlikely that it can be attributed to him) in a book written by Arleen Lorrance who was employed in a high school in Brooklyn, New York. Lorrance wrote in her book titled *The Love Project*:

> *"For seven years I served my sentence and marked off institutional time; I complained, cried, accepted hopelessness, put down the rest of the faculty for all the things they didn't do, and devoted all my energies to trying to change others and the system. It came to me loud and clear that I was the only one who could imprison (or release) me, that I was the only one I could do anything about changing. So, I let go of my anger and negativism and made a decision to simply be totally loving, open, and vulnerable all the time. Be the change you want to see happen."*[12]

Meaning in the 21st century comes, as it always has, from within each individual. Perhaps if I had just returned from a 40-day wilderness experience, or a burning bush in my backyard spoke to me, or even if I found some golden tablets with a strange language written on them as I was digging in the backyard, you would instantly want my opinion on why we are here. But, alas, such experiences have not come my way.

SOME FINAL THOUGHTS / 283

I am just a regular human like you. I make no pretenses to be inspired by any supernatural force. When I consider meaning in the 21st century, I find that meaning comes from within me after considering a wide range of possibilities.

Years ago, my wife and I visited Colombia in South America. At the time of our visit, Colombia was engaged in violent unrest throughout parts of the country due to the drug lords, FARC (the Revolutionary Armed Forces of Colombia—the People's Army), and other battling political groups. Although our trip was to a more touristy area on the coast of this country, on our second day we met others whose tour bus had been stopped by machine-gun wielding robbers of unknown political association.

None-the-less we managed to avoid any dangers although the thought of violence was in the back of our minds as we went on various excursions that included paddling through a jungle with Indigenous people to reach the ocean where we were treated to fresh-caught fish being cooked on the beach. Another time we visited the El Totumo mud Volcano where we enjoyed floating in a thick grey mud in the crater of a small volcano, while receiving a body massage from one of the attendants. When we visited this site, there was the remains of a large snake (a large venomous snake known as a bushmaster) whose head had been severed by a machete, hanging on the fence leading into the area. There were very few of us visiting the mud volcano at this time so there was definitely a feeling of being in the middle of nowhere. I would imagine that today with massive cruise ships docking in Cartagena that there is an entirely different feel to the mud bath destination.

After a wonderful week in Colombia, on return to my home country I found myself attempting to keep abreast of the ongoing attempts to stop what amounted to a civil war throughout parts of Colombia.

On February 23, 2002, Ingrid Betancourt—the founder of the Green Party in Colombia (Partido Verde Oxigeno)—was campaigning in a remote part of the country for an upcoming election when she was abducted by FARC members. Betancourt was held captive for six and a half years in various primitive jungle locations. For much of this time, she was bound to various trees with a heavy metal chain wrapped around her neck.

Often sick, terrorized by constant insects, humiliated and taunted by her captors, Betancourt was often pushed and dragged throughout the massive Amazon area jungle in order that her captors could prevent anyone from finding her. During one lengthy period of sickness, Betancourt wrote:

> *"I was drinking hardly anything, and I ate nothing. When I went to the toilet, which was constantly, a greenish, slimy liquid left my body, excruciating, and I threw up blood, more out of weariness than from some violent urge; my skin was covered with burning pustules that I scratched until they bled to stop the itching."*[13]

In addition to her physical discomfort, Betancourt's emotional suffering was tremendous. Her captors informed her, during her second month of captivity, that her father—who she was very close to—had died of a heart attack. She had no way to communicate her grief to any family members that included her two children (a boy and a girl) and her mother. For six and a half years, she wondered if she would ever see her children again.

Six and a half years into her nightmarish hell, Betancourt was rescued by the Colombian military. Just before her rescue, when it appeared to her that she might spend the remainder of her life in jungle captivity, living like an animal, she wrote:

> *"Having lost all my freedom and, with it, everything that mattered to me—my children, my mom, my life, my dreams—not able to move around, to talk, to eat and drink, to carry out my most basic body needs—subjected to constant humiliation. I still had the most important freedom of all. No one could take it away from me. That was the freedom to choose what kind of person I wanted to be."*[14]

In a sense, we all go through life with a chain wrapped around our necks and tied to a tree. It is the chain of the past that often prevent us from finding happiness in the present.

SOME FINAL THOUGHTS / 285

As a child (around the age of 8 or 9), I loved it when the circus would come to the area of the city where I lived. With a huge open field at the end of our street, once a year the circus would set up their large tents. As kids, we used to visit the circus site as they were setting up. If we helped in some small way by carrying some materials for the workers, we would be given free tickets to the circus.

I was always fascinated by the huge elephants. These massive animals had the end of a rope tied around one ankle (or sometimes their neck) while the other end of the rope was tied to a small wooden stake that had been driven into the ground. There was no doubt in my mind that any of the elephants could have easily pulled the stake out of the ground, but they never tried.

From the circus workers, I learned that when elephants are babies, they are tied to a secure pole with a piece of rope (or sometimes a chain). Although the baby elephants try to get free, the rope is too strong for them to break and the pole is to secure. Over time, the babies stop trying to get free. As adults, the elephants rarely try to break free from the rope that is holding them even though they could easily do so. They simply, and blindly, accept their fate which is based on past learning, rather than their present potential. The elephants become trapped by their past. They became captive to a belief system that was forced on them when they were babies.

One of the greatest gifts that we have as humans is the power to choose. While we can't always choose our life circumstances, we can choose how we are going to respond to anything that comes our way.

As the pace of technology increases in the 21st century, we have a choice: we can fight against change, finding comfort in being chained to the past, or we can accept change and attempt to educate ourselves on how it impacts us, finding personal meaning that is rooted in the present, while cooperatively contributing to the well-being of others.

NOTES

Introduction

1. Douglas Adams, *The Hitchhiker's Guide to the Galaxy* (New York: Del Rey Books, 1979), 9.

2. Josh Katz, "Drug Deaths in America are Rising Faster Than Ever," *New York Times*, June 6, 2017, www.nytimes.com/

3. Sabrina Tavernise, "U.S. Suicide Rate Surges to a 30-Year High," *New York Times*, Apr. 22, 2016, www.nytimes.com/

4. "Guns in the US: The Statistics behind the violence," *BBC News*, Jan. 5, 2016, www.bbc.com/

5. Martyn Whittock, *A Brief History of Life in the Middle Ages* (London: Little, Brown Book Group, 2009), 22.

6. "Monitoring health for the SDGs Annex B: tables of health statistics by country, 2016," *World Health Organization*, Dec. 1 2017, www.who.int/

7. Steven Pinker, *The Better Angels of Our Nature* (New York: Penguin Books, 2012), p xxi.

8. Steven Pinker, "The Bright Side," *Time Magazine*, Jan. 15, 2018.

9. "The State of the Church, 2016," *Barna*, Sept. 15, 2015, www.barna.com/

10. "Two-Thirds of Christians Face Doubt," *Barna*, July 25, 2017, www.barna.com/

11. "U.S. Public Becoming Less Religious," *Pew Research Center*, Nov. 3, 2015, www.pewforum.org/

12. Nina Burleigh, "Evangelical Christians Helped Elect Donald Trump, but Their Time as a Major Political Force Is Coming to an End," *Newsweek*, Dec. 13, 2018.

13. Ahmed Eld, "UnMosqued: Why Are Young Muslims Leaving American Mosques?" *HuffPost*, Dec. 13, 2013, www.huffingtonpost.com/

14. Lydia Saad, "Record Few Americans Believe Bible Is Literal Word of God," *Gallup News*, May 14, 2017, www.gallup.com/

Chapter 1 – Fake News

1. Ian Sample, "Oldest Homo sapiens bones ever found shake foundations of the human story," *The Guardian*, June 7, 2017, www.theguardian.com/

2. L.K. Delezene, W.H. Kimbel, "Lucy redux: A review of research on Australopithecus afarensis," *Yearbook of physical Anthropology, 52*, 2009, 2-28.

3. K.A. Kitchen, "The Chronology of Ancient Egypt," *World Archaeology: Chronologies*, 23, 1991, 202.

4. Lucia Gahlin, *Gods, Rites, Rituals and Religion of Ancient Egypt* (Leicestershire: Hermes House, 2011), 72-73.

5. Robert Wright, *The Evolution of God* (New York: Little, Brown and Company, 2009), 11.

6. David P. Silverman, *Ancient Egypt* (New York: Oxford University Press, 1997) 150-151.

7. Homer, Bernard Knox (Editor), *The Odyssey*, (Hertfordshire: Woodsworth Classics, 1992), Book 3, lines 4-8.

8. Lisa Randall, *Knocking On Heaven's Door* (New York: HarperCollins, 2012), 4.

9. Richard Dawkins, *The God Delusion* (New York: Mariner Books, Houghton Mifflin Harcourt, 2008), 346.

10. Thomas Vinciguerra, "The Truce of Christmas, 1914," *The New York Times*, Dec. 25, 2005.

11. Jennifer O'Mearam, "Oshawa church strips gay woman of membership," *Toronto Star*, Nov. 15, 2018.

12. Ali A. Rizvi. *The Atheist Muslim – A Journey from Religion to Reason* (New York: St. Martin's Press, 2016), 128.

13. Adam Rutherford, *A Brief History of Everyone Who Ever Lived* (New York: The Experiment, LLC, 2017), 3.

14. Michael Grant, *Herod the Great* (Winter Park, FL: American Heritage Press, 1971), 56.

15. Yuval Noah Harari, *21 Lessons for the 21st Century* (Toronto: Penguin Random House Canada Ltd., 2018), 239.

16. Philip Liebermann, *Uniquely Human* (Cambridge: Harvard University Press, 1993), 83.

17. George Kish, *A Source Book in Geography* (Cambridge: Harvard University Press, 1978), 51.

18. Kortney Hogan, "Aristarchus: The Copernicus of Antiquity," *Cosmoquest.org*, Dec. 20, 2010, www.cosmoquest.org/

19. Henry C. King, *The History of the Telescope* (Mineola, NY: Dover Books, 2011) 30.

20. Galileo Galilei, Maurice A. Finocchiaro (Editor and Translator), *Galileo on the World Systems: A New and Abridged Translation and Guide* (Berkeley: University of California Press, 1997), 1.

21. George Orwell, *1984* (Middlesex, UK: Penguin Modern Classics, 1972), 172.

22. Bernard Lazare, *Antisemitism: Its History and Causes* (Amazon Digital Services, 2014) 114-5.

23. Kevin Young, "Moon Shot: Race, a Hoax, and the Birth of Fake News," *The New Yorker*, Oct. 21, 2017.

24. Glenn Kessler, Salvador Rizzo, and Meg Kelly, "President Trump has

made 4,229 false or misleading claims in 558 days," *Washington Post*, Aug. 1, 2018.

25. Oswyn Murray. *The Oxford History of the Classical World* (Toronto: Oxford University Press, 1986), 186-203.

26. Amanda Lindhout and Sara Corbett, *A House in the Sky* (Toronto: Scribner, 2013), 177.

27. Viktor E. Frankl, *Man's Search for Meaning* (New York: Pocket Books, 1984), 126.

Chapter 2 – Why Do People Blame God for Pain and Suffering?

1. Lee Strobel, *The Case for Faith: A Journalist Investigates the Toughest Objections to Christianity* (Grand Rapids, MI: Zondervan, 2000), 29.

2. Laurie Goodstein, Terrorism in the US, *New York Times*, Sept. 19, 2001, www.nytimes.com/

3. Marni Jackson, *Pain, The Science and Culture of Why We Hurt* (Toronto: Vintage Canada, 2003), 13.

4. Thomas Aquinas, *Summa Theologica* (Cincinnati: Benziger Bros., 1948), I. 49.

5. C. S. Lewis, *The Problem of Pain* (New York: HarperCollins, 1940), 91.

6. Conrad Hackett and David McClendon, "Christians remain world's largest religious group, but they are declining in Europe," *FactTank, News in Numbers*, Apr. 5, 2017, *www.pewresearch.org/*

7. Hackett and McClendon, "Christians remain world's largest religious group, but they are declining in Europe."

8. David Hume, *Dialogues Concerning Natural Religions* (Cambridge: Hackett Publishing Co., 1980), Part X, 63.

9. Lucia Gahlin, *Gods, Rites, Rituals and Religion of Ancient Egypt* (Leicestershire, UK: Anness Publishing, 2011), 64-68.

10. Erik Hornung, *Conceptions of God in Egypt: The One and the Many* (Ithica, NY: Cornell University Press, 1982), 203-206.

11. Lucia Gahlin, *Gods, Rites, Rituals and Religion of Ancient Egypt* (Leicestershire, UK: Anness Publishing, 2011), 17.

12. Robert Garland, *Ancient Greece: Everyday Life in the Birthplace of Western Civilization* (New York: Sterling Publishing, 2013) 257.

13. Jorg Rupke, *Religion of the Romans* (Cambridge: Polity Press, 2007), 149.

14. Julien Riel-Salvatore, "Early human burials varied but most were simple," *Science Daily*, Feb. 21, 2013, www.sciencedaily.com/

15. Dimitra Papagianni & Michael A. Morse, *The Neanderthals Rediscovered: How Modern Science is Rewriting Their Story* (New York: Thames & Hudson, 2013), 116-117.

16. Darrel W. Ray, *The God Virus: How Religion Infects Our Lives and Culture* (Bonner Springs, Kansas: IPC Press, 2009), 12.

17. Erika Andersen, "10 Quotes From the 'First Lady of the World,'" *Forbes*, Jan. 10, 2013.

18. Chris Cuomo, *CNN*, TV Broadcast, Nov. 5, 2017.

19. Don Lemon, *CNN*, TV Broadcast, Nov. 7, 2017.

20. Toyofumi Ogura, *Letters from the End of the World: A Firsthand Account of the Bombing of Hiroshima* (New York: Kodansha International, 1948) 49-61.

21. Ruhollah Khomeini, *"American plots against Iran"* speech at Iranian Central Insurance Office Staff, Imam's Sahifeh, QUM, Nov. 5, 1979.

22. Kevin M. Kruse, "Billy Graham, 'America's pastor'?" *Washington Post*, Feb. 22, 2018.

23. Guenter Lewy, *America in Vietnam* (Toronto: Oxford University Press,

1978), 442-453.

24. Yuval Noah Harari, *A Brief History of Humankind* (Toronto: McClelland and Stewart, 2014), Chapter 12.

25. Robert Jean Knecht, *The Rise and Fall of Renaissance France, 1483-1610* (London: Fontana Press, 1996), 424.

26. Doris Bergen, *War and Genocide: A Concise History of the Holocaust, Third Edition* (Lanham, Maryland: Rowman and Littlefield, 2016) 14-17.

27. "Medieval Sourcebook: Twelfth Ecumenical Council," Lateran IV, 1215, Canon 68, accessed 10 Sept., 2018, www.sourcebooks.fordham.edu.basis/lateran4.asp/

28. Richard S. Levy (Editor), *Antisemitism: A Historical Encyclopedia of Prejudice and Persecution, Volume 1* (Santa Barbara: ABC/CLIO, 2005), 779.

29. Martyn Whittock, *A Brief History of Life in the Middle Ages* (London: Little, Brown Book Group, 2009), 200.

30. Frank N. Magill (Editor), *Gregory IX, Dictionary of World Biography, Vol. 2* (London: Routledge, 1998), 400 - 402.

31. Gerard S. Sloyan, "Christian Persecution of Jews over the Centuries," *United States Holocaust Memorial Museum* website, accessed Nov. 10, 2017, www.ushmm.org/

32. "Congregation for the Doctrine of the Faith – Profile," *Vatican* website, accessed Nov. 28, 2017, www.vatican.va/

33. Elphege Vacandard (Author), Bertrand Conway (Translator), *The Inquisition, A Critical and Historical Study of the Coercive Power of the Church* (Sydney: Wentworth Press, 2016), Chapter VI.

34. Cullen Murphy, *God's Jury: The Inquisition and the Making of the Modern World* (New York: Houghton Mifflin Harcourt Publishing, 2012), 22.

35. Tracy Larissas, *Torture and Brutality in Medieval Literature: Negotiations of National Identity* (Cambridge: D.S. Brewer Ltd, 2012), 21-22.

36. Alister E. McGrath, *A life of John Calvin* (Maiden, Massachusetts: Blackwell Publishers Ltd., 1990), 118-120.

37. D.M. Bennett, *The champions of the church: their crimes and persecutions* (Liberal and Scientific Publishing House, 1878, reprinted on Google ebooks), 832.

38. John L. Stoddard, *Rebuilding a Lost Faith* (Charlotte, NC: TAN Books, reprint edition, 1990), 96.

39. Gerard S. Sloyan, "*Christian Persecution of Jews over the Centuries,*" *United States Holocaust Memorial Museum* website, accessed Nov. 10, 2017, www.ushmm.org/

40. Norman H. Baynes, *The Speeches of Adolf Hitler, Vol-II* (Delhi: Gyan Books Ltd., 1942), 19-20.

41. Rev. Dr. Mae Elise Cannon, "Religious Persecution: The Tide of the 21st Century?" *HuffPost*, Nov. 30, 2014, www.huffingtonpost.com/

42. Toby Helm, "Extremist religion is at root of 21st century wars, says Tony Blair," *The Guardian*, Jan. 25, 2014, www.theguardian.com/

43. Martha Edwards, "Religion and Cancer: Some Believe God Causes Injuries and Cancer, Study Finds," *HuffPost*, Oct. 10, 2011, www.huffingtonpost.com/

44. Howard Phillips, "Why did it happen? Religious Explanations of the 'Spanish' Flu Epidemic in South Africa," *Historically Speaking*, Volume 9, Number 7, Sept/Oct. 2008, 34-36.

45. George Peter Murdock, *Our Primitive Contemporaries* (London: Macmillan, 1934), 255.

46. H. Arlo Nimmo, *Pele, Volcano Goddess of Hawai'i: A History* (Jefferson, NC: McFarland, 2011) 208.

47. Walter Burkert, *Greek Religion* (Cambridge: Harvard University Press, 1985), 136-139.

48. Dr. Erwin W. Lutzer, *"God and Natural Disasters," Moody Church Media*

website, accessed Mar. 2, 2019, www.oneplace.com/

Chapter 3 – What Does 21St Century Learning Teach Us About Pain and Suffering?

1. L.R. Jeppson, Hartford H Keifer, Edward William, *Mites Injurious to Economic Plants* (Berkeley: University of California Press, 1975), 1-3.

2. Edward E. Ruppert, Richard S. Barnes, *Invertebrate Zoology – 7th Edition* (Boston: Cengage Learning, 2003), 590-595.

3. Stephanie Pappas, "Dogs Bring Swarm of Bacteria Into Your Home," *LiveScience*, May 22, 2013, www.livescience.com/

4. Dr. Marty Becker, "Is It Safe for Cats to Be on the Counter?" *Vet Street*, June 24, 2013, www.vetstreet.com/

5. Easyology Pets, "Your Cat's Dirty Litter Box Is Putting You Both at Risk," accessed from the *Easyology Pets* website, Nov. 10, 2017, www.easyologypets.com/

6. Richard Schiffman, "Are Pets the New Probiotic?" *New York Times*, June 6, 2017, www.nytimes.com/

7. "The human being – a bacteria controlled superorgaism," Karolinska Institute, Aug. 30, 2013, www.ki.se/en/news/

8. J. Fox, "Ten Times More Microbial Cells than Body Cells in Humans?" *Microbe*, December, 2013, 494, www.asmscience.org/

9. Nicholas Bakalar, "37.2 Trillion: Galaxies of Human Cells," *The New York Times*, June 19, 2015, www.nytimes.com/

10. F. Sommer, F. Backhed, "The gut microbiota-masters of host development and physiology," *Nat. Rev. Microbiology*, 11 – 4, 2013, 227-238.

11. Richard Schiffman, "Are Pets the New Probiotic?" *New York Times*, June 6, 2017, www.nytimes.com/

12. Jack London, *The Call of the Wild* (London: MacMillan Collector's Li-

brary, 2017), 47.

13. G. A. Polis, "The Evolution and dynamics of intraspecific predation," *Annual Review of Ecology and Systematics*, 1981, 225-251.

14. G. Hausfater, S.B. Hrdy (Editors), *Infanticide: Comparative and Evolutionary Perspectives* (London: Aldine, 1984), 500-502.

15. Tia Ghose, "Why Shark Embryos Gobble Eat Each Other Up in Utero," *LiveScience*, Apr. 30, 2013, www.livescience.com/

16. Donovan Vincent, "When moms kill their young," *Toronto Star*, Sunday, Dec. 11, 2011, www.thestar.com/

17. Jeff McMahan, Ph.D., "The Meat Eaters," *New York Times*, Sept. 19, 2010, www.nytimes.com/

18. *American Heritage Dictionary, Fifth Edition* (Boston: Houghton Mifflin Harcourt Publishing, 2016).

19. Aditya Mani Jha, "Documenting the invisible hand of genetics," *The Hindu Business Line*, Aug. 5, 2016, www.thehindubusinessline.com/

20. Richard Dawkins, *The Greatest Show On Earth –The Evidence For Evolution* (New York: Free Press, 2010), 8.

21. Jimmy A. McGuire et al., "Molecular Phylogenetics and the Diversification of Hummingbirds," *Current Biology*, Vol. 24, Issue 8, Apr. 14, 2014, 910-916, www.cell.com/

22. Brian Hutchinson, "The Shark That Can Walk on Land," accessed Nov. 14, 2018, www.oceanicsociety.org/

23. Edward B. Daeschier, Neil H. Shubin and Farish A. Jenkins, Jr., "A Devonian tetrapod-like fish and the evolution of the tetrapod body plan," *Nature*, 440, Apr. 6, 2006, 757-763, www.nature.com

24. Corey Binns, "Why animals left the sea for land," *Science* on NBC News, Aug. 14, 2007, www.nbcnews.com/

25. David M. Morens, Anthony s. Fauci, "The 1918 influenza pandemic:

insights for the 21st century," *Journal of Infectious Diseases*, 2007, 24-6, 594-617, www.academic.oup.com/jid/

26. Dr. Ananya Mandal, MD, "What is a Virus?" *News, Medical Life Sciences*, Aug. 3, 2017, www.news-medical.net/

27. "Viruses that can lead to cancer," accessed Nov. 15, 2017, www.americancancersociety.com

28. Richard Dawkins, *The Selfish Gene* (New York: Oxford University Press, 2006), xxv.

29. I. Ezkurdia, D. Juan, J.M. Rodriguez, A. Frankish, M. Diekhans, J. Harrow, J. Vazquez. A. Valencia, M.L. Tress, "Multiple evidence strands suggest may be as few as 19,000 human protein-coding genes," *Human Molecular Genetics*, 2014; doi: 10.1093, www.academic.oup.com/

30. "The Human Genome Project Completion," accessed Oct. 20, 2018, www.genome.gov/

31. Carl Sagan, *Cosmos* (New York: Ballantine Books, 1985), 20.

32. Graham C.L. Davey, Ph.D., "'Spirit Possession' and Mental Health," *Psychology Today*, Dec. 31, 2014, www.pyschologytoday.com/

33. "Church Fathers: Homily 4 on Romans (Chrysostom)," accessed Dec. 2, 2017, www.newadvent.org/

34. Dean H. Hamer et al., "A linkage between DNA markers on the X chromosome and male sexual orientation," *Science*, 261, no. 5119, July 16, 1993, 321-27, www.jstor.org/

35. Leonard Mlodinow, *The Drunkard's Walk* (New York: Pantheon Books, 2008), 195.

36. Peter Brannen, "The Death of the Dinosaurs," *The New York Times*, Jan. 31, 2015, www.nytimes.com/

Chapter 4 – Are Pain and Suffering Different for Humans?

1. Donald M. Broom, "Evolution of Pain in Pain: its nature and management in man and animals," *Royal Society of Medicine International Congress Symposium Series*, 246, 2001, 2001, www.neuroscience.cam.ac.uk/

2. Christine Condo, *"Do single-celled organisms feel pain?" Quora*, Aug. 25, 2014, www.quora.com

3. Michael Pollan, "The Intelligent Plant," *The New Yorker*, Dec. 23, 2013, www.thenewyorker.com/

4. Juan Carlos Marvizon, Ph.D., "The Uniqueness of Human Suffering," *Speaking of Research*, Jan. 12, 2015, www.speakingofresearch.com/

5. A.D. Craig, "Interoception: the sense of the physiological condition of the body," Nat. Review Neuroscience, 8, Aug. 3, 2002, 655-66, www.ncbi.nlm.nih.gov/

6. Phil Mutz, "The Incredible Way Your Emotions Are Causing You Physical Pain," *HuffPost*, May 28, 2015, www.huffingtonpost.com/

7. Chris Irvine. "Emotional pain hurts more than physical pain, researchers say," *The Telegraph*, Aug. 28, 2008, www.telegraph.co.uk/

8. Alan Fogel, "Emotional and Physical Pain Activate Similar Brain Regions," *Psychology Today*, Apr. 19, 2012, www.psychologytoday.com/

9. Catherine Carrigan, *The Difference Between Pain and Suffering* (Sedona: Bikeapelli Press, 2017), 156.

10. A.L. Bauernfeind, et al., "A volumetric comparison of the insular cortex and its subregions in primates," *Journal of Human Evolution*, 64 – 4, 2013, 263-79.

11. Simon Worrall, "Yes, Animals Think and Feel. Here's How We Know," *National Geographic*, July 15, 2015, www.nationalgeographic.com/

12. Christopher Bergland, "The Neuroscience of Social Pain – Social pain activates the same brain circuitry as physical pain," *Psychology Today*, Mar. 3, 2014, www.psychologytoday.com/

13. Barbara King, *How Animals Grieve* (Chicago: University of Chicago Press, 2014), 2-3.

14. Marc Bekoff, Ph.D., "*Grief in animals: It's arrogant to think we're the only animals who mourn,*" *Psychology Today*, Oct. 29, 2009. www.psychologytoday.com/

15. Avi Selk, "Orca ends her 'tour of grief' after carrying dead calf for week," *Washington Post*, Aug. 13, 2018, www.washingtonpost.com/

16. John A. Livingston. "*Other Selves* in William Vitek and Wes Jackson, *Rooted in the land: Essays on community and place,*" (New Haven: Yale University Press, 1996) 133.

17. Carolyn Gregoire, "What Constant Exposure To Negative News is Doing To Our Mental Health," *HuffPost*, Feb. 19, 2015, www.huffingtonpost.com/

18. Peter Kinderman, "*Traumatic life events biggest cause of anxiety, depression,*" *Science News*, Oct. 16, 2013, www.sciencenews.org/

19. Jonathon Petre, "Teenagers' stress and anxiety levels are at an all-time high with middle-class children the worst affected," *The Daily Mail*, Aug. 7, 2016, www.dailymail.co.uk/

20. John Tierney, "Good News Beats Bad on Social Networks," *The New York Times*, March 18, 2013, www.nytimes.com/

21. Ray Williams, "Why We Love Bad News," *Psychology Today*, Nov. 1, 2014, www.psychologytoday.com/

22. Grace Chou Hui-Tzu, Ph.D., and Nicholas Edge, B.S., "They Are Happier and Having Better Lives than I Am – The Impact of Using Facebook on Perceptions of Others' Lives," *Cyberpsychology, Behavior, and Social Networking*, Feb., 2012, 15-2, 117-121.

23. Alice G. Walton, "Jealous Of Your Facebook Friends? Why Social Media Makes Us Bitter," *Forbes*, Jan. 22, 2013, www.forbes.com/

24. J. Bisson, S. Cosgrove, S. Lewis, NP Robert, "Post Traumatic Stress Disorder," *National Institute of Mental Health*, Feb. 2016,

www.nimh.nih.gov/

25. *Diagnostic and Statistical Manual of Mental Disorders – 5th Edition* (Arlington: American Psychiatric Publishing, 2013) 271-280.

26. "6 ways to use your mind to control pain," *Harvard Health Publishing*, Harvard Medical School, April, 2015, www.health.harvard.edu/

27. Gina Kolata, "Yes, Running Can Make You High," *The New York Times*, March 27, 2008, www.nytimes.com/

28. J.S. Beck, *Cognitive Behavior Therapy: Basics and Beyond, 2nd Ed.* (New York: The Guilford Press, 2011), 19-20.

29. Emily Singer, "Mind-Control Over Pain," *MIT Technology Review*, Dec. 19, 2005, www.technologyreview.com/

30. Clay Routledge, Ph.D., "5 Scientifically Supported Benefits of Prayer. What science can tell us about the personal and social value of prayer," *Psychology Today*, June 23, 2014, www.psychologytoday.com/

31. Benedict Carey, "Long-Awaited Medical Study Questions the Power of Prayer," *The New York Times*, March 31, 2006, www.nytimes.com/

32. Sayantani DasGupta, "The Language of Pain: Finding Words, Compassion, and Relief," assessed March 10, 2019, www.bir.med.nyu.edu/

33. Morten L. Kringelbach and Kent C. Berridge, *"The Neuroscience of Happiness and Pleasure,"* US National Library of Medicine, PMC, July 1, 2010, 659-678, www.nlm.nih.gov/

Chapter 5 – Is There Life After Death?

1. Ibn Warraq, "Vigins? What virgins?" *The Guardian*, Jan. 12, 2002, www.theguardian.com/international/

2. Ben Armstrong, *The Electric Church* (New York: Thomas Nelson, 1979), 249.

3. Dimitra Papagianni & Michael A. Morse, *The Neanderthals Rediscovered.*

(New York: Thames & Hudson, 2013), 112-113.

4. Daniella E. Bar-Yousef Mayer, Bernard Vandermeersch and Ofer Bar-Yosef, "Shells and ochre in Middle Paleolithic Qafzeh Cave, Palestine: indications for modern behavior," *Journal of Human Evolution*, Volume 56, Issue 3, March, 2009, 307-314.

5. James P. Allen, *The Essential Guide to Egyptian Mythology* (Berkley: Oxford Guide, 2003) 28.

6. Dimitra Papagianni and Michael A. Morse, *The Neanderthals Rediscovered. How Modern Science is Rewriting Their Story* (New York: Thames and Hudson, 2013), 11.

7. Melissa Healy, "As much as 2.6% of your DNA is from Neanderthals," *Los Angeles Times*, Oct. 5, 2017, www.latimes.com/

8. Katherine Harmon, "Humans interbred with Denisovans," *Scientific American*, Aug. 30, 2012, www.scientificamerican.com/

9. Peter Eardley and Carl Still, *Aquinas: A Guide for the Perplexed* (London: Continuum, 2010), 34-35.

10. Leonardo Blair, "Pope Francis Says There's a Place for Pets in Heaven, While Conservative Catholics Preach Animals Have No Souls," *The Christian Post*, Dec. 12, 2014, www.christianpost.com/

11. Plato, *Five Great Dialogues* (New York: Walter J. Black, 1942), 93.

12. Katherine Butler, "10 of the smartest animals on Earth," *MNN*, July 15, 2016, www.mnn.com/

13. Justin Gregg, "Is your toddler really smarter than a chimpanzee?" *BBC Earth*, Oct. 12, 2014, www.bbc.com/

14. Ante-Nicene Fathers, Vol. 4, 1995, 240, accessed Dec. 1, 2018, https://oll.libertyfund.org/

15. Ante-Nicene Fathers, Vol. 2, 1995, p. 245. accessed Dec. 1, 2018, https://oll.libertyfund.org/

16. Homer (author), T.E. Shaw (translator), *The Odyssey* (Hertfordshire: Wordsworth Editions), Book 11, lines 489-491.

17. Nigel Rodgers, *The Ancient Greek World* (Leiceister: Hermes House, 2010), 152.

18. Ovid (author), A.D. Melville (Translator), *The Metamorphoses* (New York: Oxford World Classics, 2009), VIII. 25.

19. Werner Jaeger, "Greeks and Jews: The First Greek Records of Jewish Religion and Civilization," *The Journal of Religion*, Vol. 18, No. 2, April, 1938, 127-143.

20. Bernardo T. Arriaza, "Beyond Death: The Chinchorro Mummies of Ancient Chile," Washington: Smithsonian Institution, 1995, assessed August 3, 1018, www.archive.org/

21. Peter Clarke, "Neuroscience, Quantum Indeterminism and the Cartesian Soul," *Brain and Cognition*, 84, 2014, 109-117 www.ncbi.nlm.nih.gov/

22. Robin Dunbar, *The Human Story: A New History of Mankind's* Evolution (London: Faber & Faber, 2004), 191.

23. Emma Badgery, "Vivid Dreams Comfort the Dying," *Scientific American*, Nov. 1, 2014, www.scientificamerican.com/

24. Charles Q. Choi, "Peace of Mind: Near-Death Experiences Now Found to Have Scientific Explanations," *Scientific American*, Sept. 12, 2011, www.scientificamerican.com/

25. Bruce Greyson, Nathan B. Fountain, Lori L. Derr, and Donna K. Broshek, "Out-of-body experiences associated with seizures," *Frontiers in Human Neuroscience*, Feb. 13, 2014, www.ncbi.nlm.nih.gov/

26. Jacqueline Ruttimann, "Are near-death experiences a dream?" *Nature*, Apr. 10, 2006, www.nature.com/

27. Dean Mobbs, Caroline Watt, *"There is nothing paranormal about near-death experiences: how neuroscience can explain seeing bright lights, meeting the dead, or being convinced you are one of them,"* Trends in Cognitive Sciences, Volume 15, Issue 10, Oct. 2011.

28. Andra M. Smith and Claude Messierwere, "Voluntary out-of-body experience: an fMRI study," *Frontiers in Human Neuroscience*, Feb. 10, 2014, www.ncbi.nlm.nih.gov/

29. Caryle Murphy, "Most Americans believe in heaven ... and hell," *Pew Research Center*, Nov. 10, 2015, www.pewresearch.org/

30. Wade Davis. *One River: Explorations and Discoveries in the Amazon Rain Forest* (New York: Simon & Schuster, 1996), 35.

Chapter 6 – Why Do Some People Resist Change?

1. David Masci, "How the Public Resolves Conflicts Between Faith and Science," *Pew Research Center*, Aug. 27, 2007, www.pewresearch.org/

2. Deepak Chopra, *The Future of God* (New York: Harmony Books, 2014), 167.

3. Ali A. Rizvi, *The Atheist Muslim – A Journey from Religion to Reason* (New York: St. Martin's Press, 2016), 29.

4. James Dobson, *Children at Risk* (New York: Word Publishing, 1990), 27.

5. Kimberly Blakerm, *The Fundamentals of Extremism: The Christian Right in America* (New Boston: New Boston Books, 2003), 8-9.

6. Bryan Mealer, "The Struggle for a New America – A Liberal's Search For God and Faith in a Divided Country," *The New Republic*, Oct. 2018, www.newrepublic.com/

7. Jerry A. Coyne, *Faith vs. Fact* (New York: Penguin Books, 2015), 83.

8. Thomas Erdbrink, "Half of Iranians say 'no' to veil law," *The New York Times*, Feb. 4, 2018, www.nytimes.com/

9. Ludovic Lado, "The Roman Catholic Church and African Religions," accessed March 2, 2019, www.theway.org.uk/

10. James Nolan, "What is it About Religion that Fosters Abuse," *ViceUK*, Aug. 5, 2015, www.vice.com/

11. Kathryn Joyce, "The Silence of the Lambs – Are Protestants concealing a Catholic-size sexual abuse scandal?" New Republic, June 20, 2017, www.newrepublic.com/

12. Julia Baird with Hayley Gleeson, 'Submit to your husbands': Women told to endure domestic violence in the name of God," *ABC News*, Aug. 10, 2017, www.abc.net.au/

13. Sarah Pulliam Balley, "Evangelical Sex Abuse Record Worse Than Catholic, Says Billy Graham's Grandson Boz Tchividijian," *HuffPost*, Oct. 2, 2013, www.huffingtonpost.com/

14. James Dobson, "Five Reasons Why Spanking Fails," accessed from Dr. James Dobson's Family Talk website on Feb. 5, 2018, www.drjamesdobson.org/

15. "The Spanking-Depression Connection," *Canadian Medical Association Journal*, No. 7, 5 Oct., 1999, 161, www.cmaj.ca/

16. Steve Chapman, "Praise the Lord, and Pass the Ammo," *Chicago Tribune*, July 1, 1999, www.chicagotribune.com/

17. "Terry Preaches Theocratic Rule 'No More Mister Nice Christian' is the Pro-Life Activist's Theme for the 90's," *The News-Sentinel*, Fort Wayne, IN, Aug.16, 1993, www.news-sentinel.com/

18. Richard Swift, "Fundamentalism: Reaching for Certainty," New Internationalist, Aug., 1990, www.newint.org/

Chapter 7 – Will We Be Able to Create Eternal Life in the 21st Century?

1. Michio Kaku, *Physics of the Future* (New York: Anchor Books, 2011), 12.

2. Arjun Walia, "Is Eternal Life Possible? One Scientist Claims It's Possible," *Collective Evolution*, Sept. 19, 2013, www.collective-evolution.com/

3. Max Roser, "Life Expectancy," accessed Feb. 17, 2019, www.ourworldindata.com/

4. David Spiegelhater, "Life expectancy: How long will you live?" *BBC Future*, Apr. 10, 2012, www.bbc.com/future/

5. Johannes Koetti, "Boundless life expectancy: The future of aging populations," *Brookings*, Mar. 23, 2016, www.brookings.edu/

6. Aubrey de Grey, *Ending Aging* (New York: St. Martin's Griffin, 2007), 8.

7. Ray Kurzweil, *The Singularity is Near* (New York: Penguin Books, 2005), 205.

8. RJ Cane, MK Borucki, "Revival and identification of bacterial spores in 25-to 40-million-year-old Dominican amber," *Science*, 258, May 19, 1995, www.science.com/

9. Stuart Mason Dambrot, "Death by Design? Spatial models show that natural selection favors genetically-limited lifespan as a lineal benefit," *PhysOrg*, July 16, 2015, www.phys.org/

10. Carol Kaesuk Yoon, "Looking Back at the Days of the Locust," *New York Times*, Apr. 23, 2002, www.nytimes.com/

11. James B. Meigs, "Inside the Future: How PopMech Predicted the Next 100 years," Popular Mechanics, Dec. 10, 2012, www.popularmechanics.com/

12. Iman Nazari, "How technology is going to change our life," *Linkedin*, Nov. 19, 2016, www.linkedin.com/in/imannazari/

13. "All About The Human Genome Project," accessed Feb. 1, 2018, www.genome.gov/

14. Angela Watercutter, "Ray Bradbury on Sci-Fi, God and Robots: The Late Author's Biggest Ideas," *Wired*, June 6, 2012, www.wired.com/

15. C. Ballard, S. Gauthier, A. Corbett, et al., "Alzheimer's disease," *Lancet*, March 19, 2011, 377, 1019-31, www.thelancet.com/

16. "Dementia Fact sheet," *WHO*, No. 362, Mar. 18, 2015, www.who.int/

17. E. Masliah, et al., "Abeta vaccination effects on plaque pathology in

the absence of encephalitis in Alzheimer disease," *Neurology*, 64 (1), 2005), 129-131, www.ncbi.nlm.nih.gov/

18. A. Dillin, DE Gottschling, T. Nystrom, "The good and the bad of being connected: the integrons of aging," *Current Opinions, Cell Biology*, 2014, 107-12, www.ncbi.nlm.nih.gov/

19. L. Partridge, NH Barton, "Optimality, mutation and the evolution of aging," *Nature*, 362, 1993, 305-311, www.nature.com/

20. OM Pereira-Smith, Y. Ning, "Molecular genetic studies of cellular senescence," *Mutation Research/DNA Aging*, Vol 256, Issues 2-6, March-Nov., 1991, 303-310, www.sciencedirect.com

21. S. Wakayama, T. Kohda, H. Obokata, M. Tokoro, C. Li, Y. Terashita, E. Mizutani, VT Nguyen, S. Kishigami, F. Ishino, and t. Wakayama, "Successful serial recloning in the mouse over multiple generations," *Cell Stem Cell*, 12, 2013, 293-297, www.cell.com/

22. Gregg Easterbrook, "What happens When We All Live to 100?" *The Atlantic*, Oct., 2014, www/theatlantic.com/

23. Katrina Brooker, "Google Ventures and the Search for Immortality," *Bloomberg*, Mar. 9, 2015, www.bloomberg.com/

24. Clare Wilson, "Everyday drugs could give extra years of life," *New Scientist*, Oct. 1,2014, www.newscientist.com/

25. DE Harrison, et al., "Rapamycin fed late in life extends lifespan in genetically heterogeneous mice," *Nature*, 460, July, 2209, 392-395, www.nature.com/

26. SL Apelo Arriola, DW Lamming, "Rapamycin: An InhibiTOR of Aging Emergess From the Soil of Easter Island," *The Journals of gerontology Series: Biological sciences and medical sciences*, 71, July, 2016, 841-9, www.ncbi.nlm.nih.gov/

27. J.M. Campbell, S.M. Bellman, M.D. Stephenson, and K. Lisy, "Metformin reduces all-cause mortality and diseases of ageing independent of its effect on diabetes control: A systematic review and meta-analysis," *Aging Research Reviews*, 40, 2017,31-44, www.ncbi.nlm.nih.gov/

28. Sarah Knapton, Science Editor, "*World's first anti-ageing drug could see humans live to 120,*" The Telegraph, Nov. 29, 2015, www.telegraph.co.uk/

29. Jacqueline Howard, "New class of drugs targets ageing to keep you healthy," *CNN*, Sept. 5, 2017, www.cnn.com/

30. "For Canada's seniors, influenza can post a serious health threat," Canadian Medical Association, accessed Feb. 27, 2019, www.demandaplan.ca/

31. Aubrey de Grey, *Ending Aging* (New York: St. Martin's Griffin, 2007), 202.

32. KJ O'Byrne, AG Dalgleish, "Chronic immune activation and inflammation as the cause of malignancy," *British Journal of Cancer*, 85, Aug. 2001, 473-83, www.nature.com/bjc/

33. Helen Briggs, "US approves genetically modified salmon," *BBC News*, Nov. 19, 2015, www.bbc.com/

34. "*Frequently asked questions on genetically modified foods,*" WHO, May, 2014, www.who.int/foodsafety/

35. "Long-term Global Perspective Studies," FAO, accessed Mar. 3, 2018, www.fao.org/

36. Josh Mitteldorf and Dorion Sagan, *Cracking the Aging Code* (New York: Flatiron Books, 2016), 266.

37. Elizabeth Fernandez, "Lifestyle Changes May Lengthen Telomeres, A Measure of Cell Aging," *Lancet Oncology*, Vol 14. No. 11, Oct., 2013, 1112-1120, www.thelancet.com/

38. Bradford S. Weeks, "FDA and honeybee products," *WeeksMD*, Apr. 5, 2010, www.weeksmd.com/

39. Ian Sample, "Harvard scientists reverse the ageing process in mice – now for humans," *The Guardian*, Nov. 28, 2010, www.the guardian.com/

40. Heidi Ledford, "CRISPR fixes disease gene in viable human embryos," *Nature*, Vol, 548, Issue 7665, Aug/ 2, 2017, www.nature.com/

41. KM Esvelt, HM Wang, "Genome-scale engineering for systems and synthetic biology," Mol Syst Biol. 9, 1, 641, 2012, www.ncbi.nlm.nih.gov/

42. Jennifer A. Doudna and Samuel H. Sternberg, *A Crack in Creation* (New York: Houghton Mifflin Harcourt Publishing Company, 2017), 117.

43. Ewen Callaway, "Destroying worn-out cells makes mice live longer," Nature, Feb. 3, 2016, www.nature.com/

44. Karen Weintraub, *"Aging is Reversible – at Least in Human Cells and Live Mice,"* Scientific American, Dec. 15, 2016, www.scientificamerican.com/

45. William Faloon, Gregory M. Fahy, George Church, "Age-Reversal Research at Harvard Medical School," *Life Extension Magazine*, July, 2016, www.lifeextension.com/

46. "FDA approval brings first gene therapy to the United States," *U.S. Food and Drug Administration News Release*, Aug. 30, 2017, www.fda.gov/

47. George Johnson, "Unearthing Prehistoric Tumors, and Debate," *The New York Times*, 28 Dec. 28, 2010, www.nytimes.com/

48. F. Mavilio, G. Ferrari, "Genetic medication of somatic stem cells. The progress, problems and prospects of a new therapeutic technology," *EMBO Reports*, 9 Suppl 1, July, 2008, 64-9, www.ncbi.nlm.nih.gov/

49. Malcolm Ritter, "Scientists use the Dolly method to clone monkey," *The Toronto Star*, Jan. 26, 2018, www.thestar.com/

50. Pallab Ghosh, "Sperm count drop 'could make humans extinct'," *BBC News: Health*, July 25, 2017, www.bbc.com.news/

51. Dana Dovey, "The Science of Human Cloning: How Far We've Come and How Far We're Capable of Going," Medical Daily, June 25, 2015, www.medicaldaily.com/

52. John Ydstie and Joe Palica, "Embryonic Stem Cells Made Without Embryos," *NPR*, Nov. 21, 2007, www.npr.org/

53. Alice Park, "George W. Bush and the Stem Cell Research Funding Ban," (*TIME*, Aug. 20, 2012, www.time.com/

54. "Organ Donation and Transplantation Statistics," National Kidney Foundation, accessed Feb. 1, 2018, www.kidney.org/

55. "Catholic Support for Ethically Acceptable Stem Cell Research," USCCB, accessed Jan. 20, 2018, www.catholicnewsagency.com/

56. Alessandro Speciale, "Vatican gets behind adult stem cell research," *National Catholic Reporter*, Apr. 9, 2013, www.ncronline.org/

57. Bryn Nelson, "How umbilical cords are saving the lives of cancer patients," *Independent*, Tuesday, Mar. 28, 2017, www.independent.co.uk/

58. Alexey Bersenev, "Human MSC extend life span in rats," *Stem Cell Assays*, Aug. 29, 2015, https://stemcellassays.com/

59. "World's First Stem-Cell Drug Approved," *Medical News Today*, May 21, 2012, www.medicalnewstoday.com/

60. Peter Diamandis, "Why Stem Cells may save your life," assessed Jan. 18, 2019, www.diamandis.com/

61. Li-Tzu Wang, et al., "Human mesenchymal stem cells (MSCs) for treatment of immune- and inflammation-mediated diseases: review of current clinical trials," *PMC: U.S. National Library of Medicine*. v 23, 2016, www.ncbi.nlm.nih.gov/

62. Sarah Knapton, "World's first human-sheep hybrids pave way for diabetes cure and mass organ transplants," *The Telegraph*, Feb. 17, 2018, www-telegraph.co.uk/

63. Antonio Regalado, "Top U.S. Intelligence Official Calls Gene Editing a WMD Threat," MIT Technology Review, assessed Jan. 30, 2017, www.technologyreview.com/

64. Rachel S. Edgar, et al., "Peroxiredoxins are conserved markers of circadian rhythms," *Nature*, 485, 459E, May 24, 2012, 459-464, www.nature.com/

65. Alison E. Berman, "How Nanotech Will Lead to a Better Future for Us All," *SingularityHub*, Aug. 12, 2016, www.singularityhub.com/

66. Theresa Phillips, "Nanomedicine and Disease," *The Balance*, Oct. 14, 2016, www.thebalance.com/

67. Olga Oskman, "How nanotechnology research could cure cancer and other diseases," The Guardian, June 11, 2016, www.theguardian.com/

68. Robert A. Freitas, Jr, "Death Is an Outrage!" (presented at the Fifth Alcor Conference on Extreme Life Extension, Newport Beach, California, Nov. 16, 2002).

69. Alok Jha, "Nanotechnology world: Nanomedicine offers new cures," *The Guardian*, Sept. 6, 2011), www.theguardian.com/

70. Swapna Upadhyay, "Wonders of Nanotechnology in the Treatment for Chronic Lung Diseases," *Journal of Nanomedicine & Nanotechnology*, 12 Nov., 2015, www.omicsonline.org/

71. Yuval Noah Harari, *Homo Deus* (Toronto: McClelland & Stewart, 2015), 22.

Chapter 8 – Will Virtual Reality Allow Us to Create Our Own Narrative?

1. Adrian Owen, *Into The Gray Zone: A Neuroscientist Explores the Border Between Life and Death* (New York: Scribner, 2017), 27.

2. Francis Crick, *The Astonishing Hypothesis* (New York: Touchstone Books, 1995), 3.

3. Jim Blascvich and Jeremy Bailenson, *Infinite Reality: The Hidden Blueprint of Our Virtual Lives* (New York: HarperCollins, 2011), 3.

4. Hussain et al., "Excessive use of massively multi-player online role-playing games: A pilot study," *International Journal of Mental Health and Addiction*, 7, 2009, 563-571.

5. Kristen Lofgren, "201 Video Game Statistics and Trends – Who's Playing What & Why?" *Big Fish Blog: Trends and Statistics*, Feb. 8, 2016, www.bigfishgames.com/

NOTES / 310

6. Michael D. Gallagher, "Essential Facts about the Computer and Video Game Industry: 2015 Sales, Demographic and Usage Data," *Entertainment Software Association*, 2016, www.theessa.com/

7. Craig Nelson, *Rocket Men: The Epic Story of the First Men on the Moon* (New York: Penguin Books, 2009), 61.

8. Jason Johnson, "How Virtual Reality Will Be Used By Doctors To Treat Patients," *Kill Screen*, Dec. 7, 2016, www.killscreen.com/

9. Adam Stone, "How Virtual Reality Is Changing Military Training," *Insights*, July 13, 2017, www.insights.samsung.com/

10. Simon Parkin, "How Virtual Reality is Helping Heal Soldiers With PTSD," NBC News, Mar. 16, 2017, www.nbcnews.com/

11. Arthur Coates, "Virtual Reality in Construction: the future," The Institution of Structural Engineers, Aug. 17, 2017, www.istructe.org/

12. Pamela J. Waterman, "Virtual Reality: A Powerful Engineering Tool," *Digital Engineering 247*, July 1, 2014, www.digitalengineering247.com/

13. Siddhi Bajaj, "5 Best Stocks to Invest in the Growth of Virtual-Reality Technology," *TheStreet*, May 23, 2016, www.thestreet.com/

14. Jeremy Bailenson, *Experience on Demand – What Virtual Reality Is, How It Works, and What It Can Do* (New York: W. W. Norton & Company, 2018), 46.

15. Jenni Ogden, Ph.D., "Why Our Minds Wander…and why we shouldn't fee guilty about it," *Psychology Today*, June 21, 2015, www.psychologytoday.com/

16. Michael Corballis, *The Wandering Mind: What the Brain Does When You're Not Looking* (Chicago: The University of Chicago Press, 2015), 9.

17. John Gaudiosi, "UN Uses Virtual Reality to Raise Awareness and Money," *Fortune*, Apr. 18, 2016, www.fortune.com/

18. Jessica L. Maples-Keller, PhD, Brian E. Bunnell, PhD, Sae-Jin Kim, BA, and Barbara O. Rothbaum, PhD, "The Use of Virtual Reality Technology in the Treatment of Anxiety and Other Psychiatric Disorders,"

Harvard Review of Psychiatry, 2017, May/Jun., 25(3) 103-113, www.ncbi.nlm.nih.gov/

19. Brenda K. Wiederhold, PhD, Kenneth Gao, BS, Camelia Sulea, MD, and Mark D. Wiederhold, MD, "Virtual Reality as a Distractive Technique in Chronic Pain Patients," *Cyberpsychol Behav Soc Netw.*, June 2, 2014, 17(6), 346-352, www.ncbi.nlm.nih.gov/

20. Rosie Wolf Williams, "How Virtual Reality Helps Older Adults," *Forbes*, Mar. 14, 2017, www.forbes.com/

21. Haydn Watters, "Is virtual reality the antidote to help depressed seniors?" *CBC News*, Jan. 26, 2017, www.cbc.ca/news/

22. Bill Freeman, "Digital Immortality – Download the Mind by 2050," *Worldhealth.net*, June 4, 2005, www.worldhealth.net/

23. Matt Drake, "Human beings on brink of achieving immortality by 2050, expert reveals," *Express*, Feb. 19, 2018, www.express.co.uk/news/science/

24. Nigel Cameron, "US Government funds virtual reality research," *Bioethics.com*, June 14, 2007, www.bioethics.com/

25. Fruzsina Eordogh, "Russian Billionaire Dmitry Itskov Plans on Becoming Immortal by 2045," *Motherboard*, May 7, 2013, www.motherboard.vice.com/

26. Robert Lawrence Kuhn, "The Singularity, Virtual Immortality and the Trouble with Consciousness," *LiveScience*, Oct. 16, 2015, www.livescience.com/

27. Sandee LaMotte, "The very real health dangers of virtual reality," *CNN*, Dec. 13, 2017, www.cnn.com/

Chapter 9 – Three Degrees Fahrenheit

1. Ashley Strickland, "Global warming is killing the Great Barrier Reef, study says," *CNN*, Apr. 18, 2018, www.cnn.com/

2. Steve Hanley, "Why White Evangelicals Don't Care About Climate Change," *Cleantechnica*, Apr. 5, 2018, www.cleantechnica.com/

3. Alexander Nazaryan, "The Man Who Sold The Earth," *Newsweek*, Feb. 16, 2018, www.newsweek.com/

4. Charles Freeman, *The Closing of the Western Mind and the Fall of Reason* (New York: Vintage, 2002), 133.

5. Peggy Des Autels, Margaret P. Battin and Larry May (eds.), *Praying for a Cure: When Medical and Religious Practices Conflict* (Lantham, Maryland: Rowman & Littlefield Publishers, 1999) 11.

6. National Research Council, *Advancing the Science of Climate Change* (Washington: National Academic Press, 2010), 1.

7. National Research Council, *Advancing the Science of Climate Change* (Washington: National Academic Press, 2010), 21-22.

8. William D. Ruckelshaus, Lee M. Thomas, William K. Reilly, and Christine Todd Whiteman, "A Republican Case for Climate Action," *The New York Times*, Aug. 1, 2013, www.nytimes.com/

9. "Global Climate Change," NASA, accessed July 23, 2018, https://climate.nasa.gov/evidence/

10. Jeffrey Bennett, Ph.D., *A Global Warming Primer* (Boulder, Colorado: Big Kid Science, 2016), 39.

11. "Global Climate Change," NASA, accessed July 23, 2018, https://climate.nasa.gov/evidence/

12. Al Gore, "Debrief," *Bloomberg Business Week*, Sept. 24, 2018, www.bloomberg.com/

13. Paul Hockenos, "Clearly, the climate crisis is upon us," *CNN*, July 31, 2018, www.cnn.com/

14. LuAnn Dahlman, "Climate Change: Global Temperature," *Climate.gov*, Sept. 11, 2017, www.climate.gov/

15. Bob Weber, "Hundreds of Canadian glaciers melting away," *Toronto Star*, July 17, 2018, www.thestar.com/

16. Jason Samenow, "Record-breaking heat hits Norway, Finland and Sweden," *Washington Post*, July 28, 2018, www.washingtonpost.com/

17. Seth Borenstein and Frank Jordans, "Science Says: Record heat, fires worsened by climate change," *The Associated Press*, July 29, 2018, www.apnews.com/

18. Tony Juniper, *What's Really Happening To Our Planet?* (New York: Penguin Random House, 2016), 11-12.

19. Elizabeth Kolbert, *The Sixth Extinction* (New York: Henry Holt and Company, 2014), inside, front jacket.

20. Somini Sengupta, "In India, sweltering heat raises alarm," *New York Times*, July 18, 2018, www.nytimes.com/

21. Euan McKirdy. "Earth at risk of becoming a 'hothouse' if tipping point reached, report warns," *CNN*, Aug. 7, 2018, www.cnn.com/

22. Bill Nye, *Undeniable – Evolution and the Science of Creation* (New York: St. Martin's Press, 2014), 3.

23. Matt Ridley, *The Evolution of Everything* (New York: HarperCollins, 2015), 3.

24. Lindsey Kratochwill, "Where Have All The Frogs Gone? New Study Illuminates Killer Fungus," *Popular Science*, Oct. 17, 2013, www.popsci.com/

25. "What's killing the honey bees? Mystery may be solved," *CBS News*, May 14, 2014, www.cbsnews.com/

26. Dave Goulson, "The Beguiling History of Bees," *Scientific American*, Apr. 25, 2014, www.scientificamerican.com/

27. Kate Kelland, "Sperm Count Dropping Western World," *Scientific American*, July 26, 2017, www.scientificamerican.com/

28. John Brockman (editor), *This Will Change Everything* (New York: Harper Perennial, 2010), 219.

29. Laura Kane and Aleksandra Sagan, "Dangers of infectious superbugs likened to climate change: ' a slow-moving tsunami'," *The Canadian Press*, July 22, 2018, www.thecanadianpress.com/

30. Yuval Noah Harari, *Homo Deus – A Brief History of Tomorrow* (Toronto: McClelland & Stewart, 2016), 73.

31. Joel Achenback, "Dinosaurs' deaths a lesson in CO2," *Toronto Star*, June 24, 2018, www.thestar.com/

32. Richard Preston, *The Hot Zone* (New York: Anchor Books, 1994), 310-311.

33. Stephan Cass, "Solar Power Will Make a Difference – Eventually," *MIT Technology Review*, Aug. 18, 2009, www.technologyreview.com/

34. Al Gore, "Debrief," *Bloomberg Business Week*, Sept. 24, 2008, www.bloomberg.com/

35. Susan Stellin, "Reducing Your Carbon Footprint," *New York Times*, Feb. 15, 2013, www.nytimes.com/

36. Diane MacEachern, "Want to Reduce the Climate Change Impact of Your House? Follow This 10-Step Checklist," *HuffPost*, Nov. 10, 2014, www.huffingtonpost.com/

37. Nicki Sanchez, "People just like you are inspiring change," accessed June 25, 2018, www.davidsuzuki.org/

38. Gryphon Adams, "How Does Changing the Light Bulbs Help the Environment," *SFGATE*, July 31, 2018, www.sfgate.com/

39. Zachary Shahan, "NASA Says: Automobiles Largest Net Climate Change Culprit," *CleanTechnica*, Feb. 23, 2010, www.cleantechnica.com/

40. Melissa Denchak, "How You Can Stop Global Warming," *NRDC*, July 17, 2017, www.nrdc.org/

41. "Energy in Europe—State of Play," European Environmental Agency, accessed Aug. 2, 2018, www.eea.europa.eu/

42. Yale School of Forestry and Environmental Studies, "The World Added 30% More Solar Energy Capacity in 2017," *YaleEnvironment360*, Mar. 19, 2018, www.e360.yale.edu/

43. Yuan Yuan, "Green Commitment," *Newsweek*, Sept. 7, 2018, www.newsweek.com/

44. Mark Chediak and Christopher Martin, "California moves to require 100% clean electricity by 2045," *Bloomberg*, Aug. 30, 2018, www.bloomberg.com/

45. Mary Kaye Schilling, "Dust to Dust," *Newsweek*, Aug. 31, 2018, www.newsweek.com/

46. Seth Borenstein, "UN report on global warming carriers life-or-death warning," *Associated Press*, Oct. 7, 2018, www.apnews.com/

47. Justin Worland, "A New Climate for Climate," *TIME*, March 31, 2019. www.time.com/

48. Suyin Haynes, "After climate strike, students want more," *TIME*, March 31, 2019. www.time.com/

Chapter 10 – Survival of the Fittest

1. Robert Axelrod, *The Evolution of Cooperation* (New York: Basic Books, 1984), xi.

2. John Hands, *Cosmo Sapiens—Human Evolution from the Origin of the Universe* (New York: Overlook Duckworth, 2015), 329.

3. Jonathan Merritt, "Was World War I a religious crusade? An interview with Philip Jenkins," *Religious News Service*, June 28, 2018, www.religionnews.com/

4. Bill Nye, *Undeniable – Evolution and the Science of Creation* (New York: St. Martin's Press, 2014), 223.

5. David Briggs, "Study: Atheists, Christians more alike than you think," *HuffPost*, Dec. 6, 2017, www.huffingtonpost.com/

6. Frans De Waal, *The Bonobo and the Atheist* (New York: W. W. Norton & Company, 2013), 20.

7. David Niose, "Misinformation and facts about secularism and religion – Even smart people perpetuate untruths about atheism and religion," *Psychology Today*, Mar. 30, 2011, www.psychologytoday.com/

8. Camille Veselka, "A Detrimental Influence: The Effect Religion Has on Laws," *Huff Post*, Dec. 6, 2011, www.huffingtonpost.com/

9. Steven Pinker, *The Better Angels of Our Nature* (New York: Penguin Books, 2011), 490.

10. Robert Axelrod, *The Evolution of Cooperation* (New York: Basic Books, 1984), 173.

11. Martin A. Nowak, "Five rules for the evolution of cooperation," *Science*, Dec. 8, 2006, 314 (5805), 1560-1563, www.ncbi.nlm.nih.gov/

12. Charles Darwin, *The Descent of Man* (Seattle: Pacific Publishing Studio, 2011), 89.

13. Tony Ashworth, *Trench Warfare 1914-1918* (London: Pan Books, 2000), 124.

14. Alejandra Guzman, "6 ways social media is changing the world," *World Economic Forum*, Apr. 7, 2016, www.weforum.org/

15. Matt Ridley, *The Evolution of Everything* (New York: HarperCollins, 2015), 314.

16. Zeynep Tufekci, *Twitter and Tear Gas—The Power and Fragility of Networked Protest* (New Haven and London: Yale University Press, 2017), vii and viii.

17. Lucy Moore, *The Thieves' Opera* (New York: Harcourt Brace & Company, 1997), 191.

18. Ronan Farrow, *War On Peace – The End of Diplomacy and the Decline of American Influence* (New York: W.W. Norton & Company, 2018), 270.

19. Anup Shah, "World Military Spending," *Global Issues*, June 30, 2013, www.globalissues.org/

20. Matt Ridley, *The Origins of Virtue, Human Instincts and the Evolution of Cooperation* (New York: Penguin Books, 1996), 175.

21. Sue Fink, The Angel City Chorale, *America's Got Talent*, Fox TV, July 24, 2018.

22. Neil Turok, *The Universe Within* (Toronto: House of Anansi Press Inc., 2012), 202-203.

Chapter 11 – Why Are We Here?

1. Sam Harris, *Waking Up: A Guide to Spirituality Without Religion* (New York: Simon & Schuster, 2014), 15.

2. Thomas Maurice, "The History of Hindostan, Vol. 2, 1798, 171," archive.org, accessed Jan. 5, 2019, www.archive.org/

3. Selina O'Grady, *And Man Created God* (New York: St. Martin's Press, 2012), 5.

4. Bart D. Ehrnan, *Did Jesus Exist? The Historical Argument for Jesus of Nazareth* (New York: HarperCollins, 2012), 208-209.

5. Deepak Chopra, *The Future Of God* (New York: Harmony Books, 2014), 1.

6. Robert Wright, *The Evolution of God* (New York: Little, Brown, and Company. 2009), 28.

7. Lawrence M. Krauss, *The Greatest Story Ever Told* (London: Simon & Schuster, 2017), 2.

8. Caryle Murphy, "Most Americans believe in heaven... and hell," Pew Research Center, Nov. 10, 2015, www.pewresearch.com/

9. Amanda Casanova, "Survey: Majority of Western Europeans do Not Believe in Heaven or Hell," *Christian Headlines*, Dec. 8, 2017,

www.christianheadlines.com/

10. Rev. Dr. John Cameron, "Even the Pope doesn't believe in hell so why should we?" *The Independent*, Mar. 31, 2018, www.independent.co.uk/

11. Elizabeth Palermo, Associate Editor, "The Origins of Religion: How Supernatural Beliefs Evolved," *LiveScience*, Oct. 5, 2015, www.livescience.com/

12. Nicole Winfield, "Pope pushes for end to death penalty," *The Associated Press*, Aug. 3, 2018, www.apnews.com/

13. Raju Mudhar and Ilya Banares, "U.S. priests accused of sex abuse in Canada," *The Toronto Star*, August 19, 2018, www.thestar.com/

14. Ashok Sharma, "India's top court ends ban against women in temple," *The Associated Press*, Oct. 1, 2018, www.apnews.com/

15. Robyn Faith Walsh, "Is Star Wars a Religion?" *HuffPost*, Dec. 6, 2017, www.huffingtonpost.com/

16. Denyse O'Leary, *"Post-Modern Physics: String Theory Gets Over The Need for Evidence,"* *Evolution News*, July 19, 2017, https://evolutionnews.org/

17. Gerald L. Schroeder, *The Hidden Face of God – Science Reveals the Ultimate Truth* (New York: Simon & Schuster, 2001), xi.

18. Kathryn Blaze Carlson, "Organized religion on the decline? Growing number of Canadians 'spiritual but not religious'," *National Post*, Dec. 21, 2012, www.nationalpost.com/

19. Phil Zuckerman, Ph.D., "How Many Atheists Are There?" *Psychology Today*, Oct. 20, 2015, www.psychology.com/

20. Jess Stufenberg, "The six countries in the world with the most 'convinced atheists'," *The Independent*, Mar. 23, 2016, www.independent.co.uk/

21. Gabe Bullard, "The World's Newest Major Religion: No Religion," *National Geographic*, Apr. 22, 2016, www.nationalgeographic.com/

22. David Niose, "Atheism, Meaning, and the Absurdity of It All," *Psychology Today*, Feb. 16, 2014, www.psychologytoday.com/

23. Adam Lee, "Are we ready to face death without religion?" *The Guardian*, Mar. 13, 2016, www.theguardian.com/

24. Derek Beres, "New Study Reveals Which People Fear Death the Least," *Big Think*, Aug. 14, 2018, www.bigthink.com/

25. Stephen Hawking, *Brief Answers to the Big Questions* (New York: Bantam Books, 2018), 38.

26. Robert Evans, "Atheists face death in 13 countries, global discrimination: study," *Reuters*, Dec. 9, 2013, www.reuters.com/

27. Graham Phillips, "It may be possible to send life to other planets – but should we?" *The Sydney Morning Herald*, Jan. 5, 2018, www.smh.com.au/

28. Julian Baggini, "We are simulations living in a virtual realm, says Elon Musk. But why do we like the idea?" *The Guardian*, June 3, 2016, www.theguardian.com/

29. Rene Descartes, *Meditations on First Philosophy* (1641), Meditation vi.

30. Plato, *Republic* (approximately 380 BC), 514a-520a.

31. Valerie Reimer, "Ignition 2016: Virtual Reality is Here," Business Insider, Sept. 12, 2016, www.businessinsider.com/

32. Juliana Baggini, "We are simulations living in a virtual realm, says Elon Musk. But why do we like the idea?" *The Guardian*, June 3, 2016, www.theguardian.com/

33. Brian Greene, *The Fabric of the Cosmos – Space, Time, and the Texture of Reality* (New York: Random House, 2004), ix.

34. Matt Williams, "A Universe of 10 Dimensions," *Universe Today*, Nov. 7, 2016, www.universetoday.com/

35. Byron Reese, *The Fourth Age* (New York: Atria Books, 2018), 181.

Chapter 12 – Some Final Thoughts

1. UNESCO Information Services, accessed March 2, 2019, www.unesco.org/

2. Rupert Neate, "World's richest 500 see their wealth increase by $1tn this year," The Guardian, Dec. 27, 2017, www.theguardian.com/

3. Kimberly Palmer, "Do Rich People Live Longer?" U.S. News, Feb. 14, 2012, www.usnews.com/

4. Anna Hodgekiss, "Smoking cannabis really DOES make people lazy because it affects the area of the brain responsible for motivation," Daily Mail, Nov 8, 2018, www.dailymail.co.uk/

5. Emily Jashinsky, "Michael Moore blames 'dumbed-down electorate' for Trump," Washington Examiner, Nov. 8, 2018, www.washingtonexaminer.com/

6. Ian Duncan Smith, "HL Mencken predicted a moron in the White House," *The Guardian*, Oct. 13, 2017, www.theguardian.com/

7. Juan Enriquez and Steve Gullans, *Evolving Ourselves – How Unnatural Selection and Nonrandom Mutation Are Changing Life on Earth* (New York: Penguin Books, 2015), 16.

8. "Forced Sterilization," United States Holocaust Memorial Museum website, accessed Mar. 3, 2019, www.ushmm.org/

9. "Eugenics," Canada's Human Rights History, accessed Mar. 3, 2019, https://historyofrights.ca

10. "Forced Sterilization," United States Holocaust Memorial Museum website, accessed Mar.3, 2019, www.ushmm.org/

11. Peter Ward, "What Will Become of Homo Sapiens?" *Scientific American*, Jan., 2009, 68-73, www.scientificamerican.com/

12. Richard Dean Kellough (editor), *Developing Practices and a Style: Selected Readings in Education for Teachers and Parents, Second Edition* (New York: MSS Information Corporation, 1974), 85.

13. Ingrid Betancourt, *Even Silence Has An End* (London: Penguin Books, 2010), 484.

14. Ingrid Betancourt, *Even Silence Has An End* (London: Penguin Books, 2010), 491.

Suggested Readings

Adams, Douglas. *The Hitchhiker's Guide to the Galaxy*. New York: Del Rey Books, 1979.
Ali, Ayaan Hirsi. *Infidel*. New York: Free Press, 2007.
Allen, James P. *The Essential Guide to Egyptian Mythology*. Berkley: Oxford Guide, 2003.
Aquinas, Thomas. *Summa Theologica*. Cincinnati: Benziger Bros., 1948.
Armstrong, Ben. *The Electric Church*. New York: Thomas Nelson, 1979.
Armstrong, Karen. Holy War - The Crusades and Their Impact on Today's World. New York: Anchor Books, 2001.
Ashworth, Tony. *Trench Warfare*, 1914-1918. London: Pan Books, 2000.
Axelrod, Robert. *The Evolution of Cooperation*. New York: Basic Books, 1984.
Baigent, Michael. The Jesus Papers. New York: HarperCollins, 2007.
Bailenson, Jeremy. *Experience on Demand - What Virtual Reality Is, How It Works, and What It Can Do*. New York: W.W. Norton & Company, 2018.
Baynes, Norman H. The Speeches of Adolf Hitler, Vol. II. Delhi: Gyan Books Ltd., 1942.
Betancourt, Ingrid. *Even Silence Has An End*. London: Penguin Books, 2010.
Beck, J.S. *Cognitive Behavior Therapy: Basics and Beyond, 2nd Edition*. New York: The Guilford Press, 2011.
Bennett, Jeffrey. A Global Warming Primer. Boulder, Colorado: Big Kid Science, 2016.
Bergen, Doris. *War and Genocide: A Concise History of the Holocaust, Third Edition*. Lanham, Maryland: Rowman & Littlefield, 2016.
Blaker, Kimberly. *The Fundamentals of Extremism: The Christian Right in America*. New Boston: New Boston Books, 2003.
Blascvich, Jim, and Bailenson, Jeremy. *Infinite Reality: The Hidden Blueprint of Our Virtual Lives*. New York: HarperCollins, 2011.
Bradbury, Ray. *Fahrenheit 451*. New York: Simon & Schuster Reissue, 2012).
Brockman, John (Editor). *This Will Change Everything*. New York: Harper Perennial, 2010.
Burkert, Walter. *Greek Religion*. Cambridge: Harvard University Press, 1985.
Carrigan, Catherine. *The Difference Between Pain and Suffering*. Sedona: Bikeapelli Press, 2017.

READINGS / 324

Chopra, Deepak. *The Future of God*. New York: Harmony Books, 2014.
Clegg, Brian. *Ten Billion Tomorrows - How Science Fiction Technology Became Reality and Shapes the Future*. New York: St. Martin's Press, 2015.
Corballis, Michael. *The Wandering Mind: What the Brain Does When You're Not Looking*. Chicago: The University of Chicago Press, 2015.
Coyne, Jerry A. *Faith vs. Fact*. New York: Penguin Books, 2015.
Crick, Francis. *The Astonishing Hypothesis*. New York: Touchstone Books, 1995.
Darwin, Charles. *The Descent of Man*. Seattle: Pacific Publishing Studio, 2011.
Davis, Wade. *One River Explorations and Discoveries in the Amazon Rain Forest*. (New York: Simon & Schuster, 1996.
Dawkins, Richard. *The God Delusion*. New York: Houghton Mifflin Harcourt, 2008.
Dawkins, Richard. The Greatest Show on Earth - *The Evidence for Evolution*. New York: Free Press, 2010.
Dawkins, Richard. *The Selfish Gene*. New York: Oxford University Press, 2006.
De Grey, Aubrey. *Ending Aging*. New York: St. Martin's Griffin, 2007.
De Wal, Frans. *The Bonobo and the Atheist*. New York: W.W. Norton & Company, 2013.
DesAutels, Peggy, Battin, Margaret P., and May Larry (Editors). *Praying for a Cure: When Medical and Religious Practices Conflict*. Lantham, Maryland: Rowman & Littlefield Publishers, 1999.
Dobson, James. *Children at Risk*. New York: Word Publishing, 1990.\
Doudna, Jennifer A., Sternberg, Samuel H. *A Crack in Creation*. New York: Houghton Mifflin Harcourt Publishing Co., 2017.
Dunbar, Robin. *The Human Story: A New History of Mankind's Evolution*. London: Faber & Faber, 2004.
Eardley, Peter and Still, Carl. *Aquinas: A Guide for the Perplexed*. London: Continuum, 2010.
Ehrnan, Bart D. *Did Jesus Exist? The Historical Argument for Jesus of Nazareth*. New York: HarperCollins, 2012.
Eisenstein, Charles. Climate: A New Story. Berkeley: North Atlantic Books, 2018.
Enriquez, Juan, and Gullans, Steve. *Evolving Ourselves - How Unnatural Selection and Nonrandom Mutation Are Changing Life on Earth*. New York: Penguin Books, 2015.
Farrow, Ronan. *War on Peace - The of Diplomacy and the Decline of American Influence*. New York: W.W. Norton & Company, 2018.
Frankl, Viktor E. *Man's Search for Meaning*. New York: Pocket Books, 1984.

READINGS / 325

Franzen, Jonathan. *The End of the End of the Earth*. Toronto: Penguin Random House Canada, 2018.

Freeman, Charles. *The Closing of the Western Mind and the Fall of Reason*. New York: Vintage, 2002.

Gahlin, Lucia. *Gods, Rites, Rituals and Religions of Ancient Egypt*. Leicestershire: Hermes House, 2011.

Galileo Galilei, Maurice A. Finocchiaro (Editor and Translator). *Galileo on the World Systems: A New and Abridged Translation and Guide*. Berkeley: University of California Press, 1997.

Garland, Robert. *Ancient Greece: Everyday Life in the Birthplace of Western Civilization*. New York; Sterling Publishing, 2013.

Gaynor, Mitchell L. *The Gene Therapy Plan*. New York: Viking, 2015.

Gould, Stephen Jay. *Ever Since Darwin - Reflections in Natural History*. London: Burnett Books Ltd., 1978.

Grant, Michael. *Herod the Great*. Winter Park, FL: American Heritage Press, 1971.

Green, Brian. *The Future of the Cosmos - Space, Time, and the Texture of Reality*. New York: Random House, 2004.

Hands, John. *Cosmo Sapiens - Human Evolution from the Origin of the Universe*. New York: Overlook Duckworth, 329.

Harari, Yuval Noah. *A Brief History of Humankind*. Toronto; McClelland & Stewart, 2014.

Harari, Yuval Noah. *A Brief History of Tomorrow*. Toronto; McClelland & Stewart, 2016.

Harari, Yuval Noah. *Homo Deus*. Toronto: McClelland & Stewart, 2015.

Harari, Yuval Noah. *21 Lessons for the 221st Century*. Toronto; Penguin Random House Canada, 2018.

Harpur, Tom. *The Pagan Christ*. Toronto: Thomas Allen Publishers, 2004.

Harris, Sam. *Waking Up - A Guide to Spirituality Without Religion*. New York: Simon & Schuster, 2014.

Hausfater, G., Hrdy, S.B. (Editors). *Infanticide: Comparative and Evolutionary Perspectives*. London: Aldine, 1984.

Hawking, Stephen. Brief Answers to the Big Questions. New York: Bantam Books, 2018.

Hedges, Chris. *American Fascists - The Christian Right and the War on America*. New York: Free Press, 2006.

Homer (Author), Knox, Bernard (Editor). *The Odyssey*. Hertfordshire: Woodsworth Classics, 1992.

Horning, Erik. *Conceptions of God in Egypt*. Ithica, NY: Cornell University Press, 1982.

Hume, David. *Dialogues Concerning Natural Religions*. Cambridge: Hack-

ett Publishing Co., 1980.
Huxley, Aldous. *Brave New World*. New York: Harper Perennial Reprint, 2006.
Jackson, Marni. *Pain, the Science and Culture of Why We Hurt*. Toronto; Vintage Canada, 2003.
Jeppson, L.R., Keifer, Hartford H., William, Edward. *Mites Injurious to Economic Plants*. Berkeley: University of California Press, 1975.
Juniper, Tony. *What's Really Happening to Our Planet?* New York: Penguin Random House, 2016.
Kaku, Michio. *Physics of the Future*. New York; Anchor Books, 2011.
Kellough, Richard, Dean (Editor). *Developing Practices and a Style: Selected Readings in Education for Teachers and Parents, Second Edition*. New York: MSS Information Corporation, 1974.
King, Barbara. *How Animals Grieve*. Chicago: University of Chicago Press, 2014.
King, Henry C. *The History of the Telescope*. Mineola, NY: Dover Books, 2011.
Kisak, Paul F. (Editor). *CRISPR Technology: The Revolutionary Breakthrough For Genetics & Evolution*. CreateSpace Publishing, 2017.
Kish, George. *A Source Book in Geography*. Cambridge: Harvard University Press, 1978.
Klein, Naomi. *This Changes Everything*. New York: Penguin Random House, 2014.
Knecht, Robert Jean. *The Rise and Fall of Renaissance France, 1483-1610*. London: Fontana Press, 1996.
Kolbert, Elizabeth. *The Sixth Extinction*. New York: Henry Holt & Company, 2014.
Krauss, Lawrence M. *The Greatest Story Ever Told*. London: Simon % Schuster, 2017.
Kuhn, Alvin Boyd. Shadow of the Third Century: A Revaluation of Christianity. Whitefish, Montana: Kessinger Publishing, 2010.
Kurzweil, Ray. *The Singularity is Near*. New York: Penguin Books, 2005.
Lanier, Jaron. *Dawn of the New Everything - Encounters with Reality and Virtual Reality*. New York; Henry Holt and Co., 2017.
Larissas, Tracy. *Torture and Brutality in Medieval Literature: Negotiations of National Identity*. Cambridge: D.S. Brewer Ltd, 2012.
Lewis, C.S. *The Problem of Pain*. New York: HarperCollins, 1940.
Lewy, Guenter. *America in Vietnam*. Toronto: Oxford University Press, 1978.
Levy, Richard (Editor). *Antisemitism: A Historical Encyclopedia of Prejudice and Persecution, Vol. 1*. Santa Barbara: ABC/CLIO, 2005.

Liebermann, Philip. *Uniquely Human*. Cambridge: Harvard University Press, 2993.
Lindhout, Amanda, and Corbett, Sara. *A House in the Sky*. Toronto; Scribner, 2013.
Livingston, John A. *Other Selves* in William Vitek and Wes Jackson, *Rooted in the land: Essays on community and place*. New Haven: Yale University Press, 1996.
London, Jack. *The Call of the Wild*. London: MacMillan Collector's Library, 2017.
Losos, Jonathan B. *Improbable Destinies - Fate, Chance, and the Future of Evolution*. New York: Riverhead Books, 2017.
McGrath, Alister E. *A Life of John Calvin*. Maiden, Massachusetts: Blackwell Publishers Ltd., 1990.
Magill, Frank N. (Editor). *Gregory IX, Dictionary of World Biography*, Vol. 2. London: Routledge, 1998.
Michaels, Sean. UsConductors. Toronto: Vintage Canada, 2015.
Mighton, John. The End of Ignorance - Multiplying Our Human Potential. Toronto: Vintage Press, 2006.
Miller, Matt. *The Tyranny of Dead Ideas*. New York: Henry Holt and Co., 2009.
Mitteldorf, Josh, and Sagan, Dorion. *Cracking the Aging Code*. New York: Flatiron Books, 2016.
Mlodinow, Leonard. *The Drunkard's Walk*. New York: Pantheon Books, 2008.
Moore, Lucy. *The Thieves' Opera*. New York: Harcourt Brace & Company, 1997.
Murdock, George Peter. *Our Primitive Contemporaries*. London: Macmillan, 1934.
Murphy, *God's Jury*. New York: Houghton Mifflin Harcourt Publishing, 2012.
Murray, Oswyn. *The Oxford History of the Classical World*. Toronto; Oxford University Press, 1986.
Myers, William. Bio Design - Nature, Science, Creativity. New York: MOMA, 2018.
National Research Council. *Advancing the Science of Climate Change*. Washington: National Academic Press, 2010.
Nelson, Craig. *Rocket Men: The Epic Story of the First Men on the Moon*. New York: Penguin Books, 2009.
Nimmo, H. Arlo. Pele, *Volcano Goddess of Hawai'i: A History*. Jefferson, NC: McFarland, 2011.
Nye, Bill. *Undeniable - Evolution and the Science of Creation*. New York: St.

Martin's Press, 2014.
O'Grady, Selina. *And Man Created God.* New York: St. Martin's Press, 2012.
Ogura, Totofumi. *Letters from the End of the World: A Firsthand Account of the Bombing of Hiroshima.* New York: Kodansha International, 1948.
Orwell, George. *1984.* Middlesex, UK: Penguin Modern Classics, 1972.
Ovid (Author), Melville A.D. (Translator). *The Metamorphoses.* New York: Oxford World Classics, 2009.
Ovink, Henk, and Boeijenga, Jelte. *Too Big: Rebuild by Design: A Transformative Approach to Climate Change.* Rotterdam: Naioio Publishers, 2018.
Owen, Adrian. *Into the Gray Zone.- A Neuroscientist Explores the Border Between Life and Death.* New York: Scribner, 2017,
Papagianni, Dimitra, & Morse, Michael A. *The Neanderthals Rediscovered: How Modern Science is Rewriting Their Story.* New York: Thames & Hudson, 2013.
Pinker, Steven. *The Better Angels of Our Nature.* New York: Penguin Books, 2012.
Plato. *Five Great Dialogues.* New York: Walter J. Black, 1942.
Preston, Richard. *The Hot Zone.* New York: Anchor Books, 1994.
Randall, Lisa. *Knocking on Heaven's Door.* New York: Harper Collins, 2012.
Ray, Darrel W. The God Virus: How Religion Infects Our Lives and Culture. Bonner Springs, Kansas: IPC Press, 2009.
Reese, Byron. *The Fourth Age.* New York: Atria Books, 2018.
Ridley, Matt. *The Evolution of Everything.* New York: HarperCollins, 2013.
Ridley, Matt. *The Origins of Virtue, Human Instincts and the Evolution of Cooperation.* New York: Penguin Books, 1996.
Rizvi, Ali A. *The Atheist Muslim - A Journey from Religion to Reason.* New York: St. Martin's Press, 2016.
Rodgers, Nigel. *The Ancient Greek World.* Leiceister: Hermes House, 2010.
Rupke, Jorg. *Religion of the Roman.* Cambridge Polity Press, 2007.
Ruppert, Edward E., Barnes, Richard S. *Invertebrate Zoology - 7th Edition.* Boston: Cengage Learning, 2003.
Rutherford, Adam. *A Brief History of Everyone Who Ever Lived.* New York: The Experiment, LLC, 2017.
Sagan, Carl. *Cosmos.* New York: Ballantine Books, 1985.
Schroeder, Gerald L. *The Hidden Face of God - Science Reveals the Ultimate Truth.* New York: Simon & Schuster, 2001.
Silverman, David P. *Ancient Egypt.* New York: Oxford University Press, 1997.
Spencer, Robert. *The Complete Infidel's Guide to the Koran.* Washington: Regnery Publishing, 2009.
Stoddard, John L. *Rebuilding a Lost Faith.* Charlotte, NC: TAN Books,

1990.
Strobel, Lee. *The Case for Faith: A Journalist Investigates the Toughest Objections to Christianity.* Grand Rapids, MI: Zondervan, 2009.
Tudge, Colin. The Time Before History. New York: Scribner, 1996.
Tufekci, Zeynep. *Twitter and Tear Gas - The Power and Fragility of Networked Protest.* New Haven and London: Yale University Press, 2017.
Turok, Neil. *The Universe Within.* Toronto; House of Anansi Press Inc., 2012.
Vacandard, Elphege (Author), Conway, Bertrand (Translator). *The Inquisition, A Critical and Historical Study of the Coercive Power of the Church.* Sydney: Wentworth Press, 2016.
Vance, Ashlee. *Elon Musk.* New York: HarperCollins, 2015.
Wall, Patrick. Pain: The Science of Suffering. New York: Columbia University Press, 2000.
Whittock, Martyn. *A Brief History of Life in the Middle Ages.* London: Little, Brown Book Group, 2009.
Williamson, Kevin D. *The End is Near and It's Going To Be Awesome.* New York: Broadside Books, 2013.
Wilson, David Sloan. *Darwin's Cathedral - Evolution, Religion, and the Nature of Society.* Chicago: The University of Chicago Press, 2003.
Wright, Robin. *The Evolution of God.* New York: Little, Brown and Company, 2009.

About the Author

Brian Harris lectures at Queen's University. He holds degrees from Laurier and the University of Toronto. Brian has worked as a professional counselor in a wide range of educational settings including more than a decade as an emergency response crisis counsellor. He is an Amazon bestselling author of more than 15 books. Brian is also an accomplished artist (www.bcharris.com).

Also by B.C. Harris

Lost Worlds

Fiery Illusions

Final Quest

www.ingramcontent.com/pod-product-compliance
Lightning Source LLC
Chambersburg PA
CBHW051304220526
45468CB00004B/1200